landscape
architecture
风 景 园 林

吉典文化和千朋万友 编
郑 波 主编
顾玉梅 译

1

大连理工大学出版社

景观设计学（Landscape Architecture）在国外是与建筑学、城市规划相提并论的，而不是从属的学科。

该专业无论在专业起源、学科内容以及位置诸多方面都与国内的风景园林专业很相似。由于文化背景、经济发展速度、社会制度等众多因素不同，国内外园林专业发展水平和侧重都会有所差别。尽管相比之下，国外一些先进国家的景观设计学科专业面更宽、涉足的内容更多，与其他学科渗透的程度更深，但是，专业的性质是相同的，产生和发展的脉络是一致的。对于其间的差距认识应该是国内园林学科建设和努力学习的目标。

本书汇集了全球众多的景观设计案例，通过大量的照片、文字及分析图将获奖项目原汁原味地展现出来，内容覆盖风景园林实践的各个范围，包括：校园环境、乡村庄园、庭园、传统园林，居住环境、单位园林、景观规划、自然风景、城市开放空间国家森林、国家公园、新城与规划社区、风景车道、娱乐区、自然景观恢复、街景与广场、城市公园、水滨等。按国内设计师的阅读方式整编为六本：1 公园、绿地的景观规划和设计；2 广场的景观设计；3 公共庭院设计、私家庭院设计；4 地球之肺——湿地；5 历史景观保护、改造的规划和设计；6 风景旅游区、度假区的景观规划和设计。案例种类丰富新颖、前沿信息量大，便于读者全方位地参考和解读景观方案的全貌，从这些全球顶级景观新作中获取设计精髓，受益匪浅。

运动公园景观 SPORTS PARK LANDSCAPE

市政公园景观 MUNICIPAL PARK LANDSCAPE

目 录 CONTENTS

社区公园景观 COMMUNITY PARK LANDSCAPE

运动公园景观
SPORTS PARK LANDSCAPE

伯格豪森的巴伐利亚州园林运动公园
Bavarian State Garden Exhibition Burghausen Sports Park

LOCATION:
Bavarian German
AREA:
7.5 ha
TYPE:
Park
DESIGN COMPANY:
Rehwaldt Landschaftsarchitekten

项目地点:
德国 巴伐利亚州
面积:
7.5 公顷
设计公司:
雷瓦德景观建筑事务所

Within Burghausen's quarter Neustadt an urban park has been developed as the core area for the "Landesgartenschau". Right in the center of the lively quarter the relocation of a municipal building yard opened up the opportunity to completely redesign the urban open space. Around an extensive meadow differentiated garden areas have been developed. Their characteristics are corresponding to the neighbouring residential patterns and each garden features an independent formal special city.

在伯格豪森市的新城新建成了一座城市公园，这座公园也是巴伐利亚州园林展的核心区。这个崭新的城市建筑园区恰位于生气勃勃的新城，为全面重塑城市公共空间提供了契机。宽阔的草坪上隔出若干区域，分建不同的花园。每座花园都与周边的住宅特色相呼应，都是一个独立、特别的"城市"。

For the first time an interconnected urban open space system has been developed through the "Landesgartenschau". The City Park becomes an icon for the entire City of Burghausen. New walkway connections and view axes offer a hitherto unknown city experience. Consequently, the exhibition's activities are a crucial contribution to long term urban development and creation of the city image.

此次巴伐利亚州园林展首次形成了一种相互关联的城市公共空间体系。城市公园成为整个伯格豪森市的标志。新建的步行道和观景轴会引领行人看到新鲜的、前所未见的城市风景。因而，此次园林展的各项活动对城市的长远发展和树立新的城市形象都意义重大。

FredericiaC 临时运动公园
FredericiaC Temporary Sports Park

LOCATION:
Fredericia

AREA:
14 ha

ARCHITECT:
SLA

DESIGN COMPANY:
SLA

项目地点：
腓特烈西亚

面积：
14 公顷

建筑师：
SLA

设计公司：
SLA

Throughout the 20th century, Fredericia's harbour was dominated by heavy industry. This is now starting to change. In 2004, one of the harbour's most prominent industries, Kemira Grow-How, ended operations and demolition began. Realdania (a foundation working to improve the built environment) bought the Kemira site in June 2008, together with other sites along the harbour. Realdania and the city of Fredericia have since formed FredericiaC P/S in a joint venture to regenerate the harbour area. SLA was hired to help this long regeneration process along. Thus the concept of the "temporary park" was born.

在整个 20 世纪，腓特烈西亚港口是以重工业为主导产业的。现在，这里开始发生变化。2004 年，Kemira Grow-How，港口最突出的产业之一，终止了运营业务，开始进行拆迁。Realdania（用以改善建筑环境基础工作的公司）于 2008 年 6 月购买了 Kemira 及其他一些沿海港的区域。Realdania 和腓特烈西亚城联合成立了一家合资企业 FredericiaC P /S，用以重建和净化海港区域，并聘请 SLA 参与此项长期的重建过程。"临时公园"的概念由此而生了。

Since the summer of 2009 a part of the harbour had been opened to the public and had been used for everything from picnics and fishing to events such as concerts, theatre, and vintage car shows. By spring 2010, the Kemira site had been completely cleared, and construction work for the temporary park project went underway. The project was completed in phases, scheduled to open in July, August and September 2010.

自 2009 年的夏天以来，港口的一部分已经向公众开放，并且已用于从野餐到钓鱼等一系列活动，如剧院、音乐会、老式汽车展等。直到 2010 年春，Kemira 区域已经得到了全面的净化治理，此外，临时公园项目的建设工作也正在进行中。项目将分阶段完成，预计于 2010 年七、八、九月开放。

运动公园景观
SPORTS PARK LANDSCAPE

The idea behind the project is to establish a robust framework, which can subsequently be allocated different functions. The design of the framework is based on historical maps of the area. The physical expressions of the frames are determined by planting, materials, lighting, and the juxtaposition of these elements. The expression is raw and simple, as it should be easy to build and remove the temporary arrangement as the new urban area takes shape.

该项目背后的构想是要建立一个强健的体系框架，随后可以配置各种不同的功能。该框架的设计是基于该地区的历史地图的。构架的有形表现形式是由花圃、材料、照明以及这些元素的并置搭配所决定的。这些表现形式是简朴而无雕琢的，而且正因为基于刚成形的新城区，这里应该方便建设和移除临时的布局安排。

The framework consists of various temporary pavings and spatial elements which create great opportunities for a range of different activities regarding "health and exercise". The included elements are based on a great involvement of the citizen of Fredericia. Spaces have also been left open within the framework so that these can be filled in over time as new needs and wishes arise.

该体系框架是由各种各样的临时路径和空间元素组成的。这些路径和空间元素能够为一系列与"健康和运动"相关的活动创造非常多的机会。这里有多种不同的与"健康和运动"有关的主题活动项目。其中所运用的想法和元素都是基于腓特烈西亚市民大力支持和积极参与的基础之上的。因为体系框架内的空间也都是处于开放的状态的，所以随着时间推移而产生的新需求和新愿望也将会添入到整个建设中去。

海尔兹利亚运动公园
Herzeliya Sports Park

LOCATION:
Herzeliya Tel Aviv Metropolitan Area Israe
AREA:
stage A- 40 acres out of 180 acres, stage B is
currently under construction
DESIGN UNITS:
Shlomo Aronson Architects

项目地点：
以色列特拉维夫荷兹利亚

面积：
总 180 英亩，A 阶段 40 英亩，B 阶段目前正在施工

设计单位：
Shlomo Aronson 建筑师事务所

This city park successfully combines ecological necessity with functional diversity. It
provides generous recreational spaces for everyone from toddlers to retirees. The
park is watered by recycled water which made the use of much-desired lawn areas
sustainable. On the other hand, the contiguous winter flood basin has been preserved,
supporting local and transient birdlife.

这个城市公园将生态环境的必要性与功能的多样性完美地结合起来。该城市公
园为每个人，无论是小孩还是老人，都提供了足够的空间，用以进行休闲娱乐
活动。一方面，这个公园通过循环水系统来浇灌草坪，使草坪生长得郁郁葱葱，
并让茂盛的草坪持续发展生长下去；另一方面，相邻的冬季池塘区域已经被保
护了起来，这个池塘被用来供养本地的鸟群和冬季迁徙来的候鸟鸟群。

运动公园景观
SPORTS PARK LANDSCAPE

The presented project is the first stage of a large future urban park (162ha out of 72.8ha). As part of a master plan that outlines a programmatic layout and development strategy for the entire site, the first stage was planned as a cornerstone. It creates an active and intensively developed park section that addresses the first wave of demands of its projected users. At the same time it protects and exposes the rich natural processes and ecologies existing on the site. It is also intended to raise public interest and political support for the protection of the bordering winter ponds and their rich wildlife, until now hidden from view and widely snubbed as mosquito breeding grounds. Adjacent to the existing soccer stadium and the city's sports center, the park becomes an integral part of the larger recreational complex of Herzeliya.

现阶段所呈现出来的项目规划是未来整个大型城市公园规划中的第一期（第一期为16.2公顷，整体规划为72.8公顷）。作为总体项目规划的一部分，一期项目规划如同基石般重要。一期项目规划将描绘出一个纲领性布局的轮廓，并且为整个现场工程提供发展性的战略方案。这里解决了第一批潜在用户的需求，创建了一片生机勃勃的，并且具有发展潜力的公园带。与此同时，这个城市公园保护并展现了丰富的自然进程和本地的生态环境。这也是为了提高市民保护与公园接壤的冬季池塘和丰富的野生动物这一公共利益的意识、兴趣，并获得更多的政治支持；直到现在为止，这片区域还是不为人所见的，仅仅被当作是蚊子的滋生地，而被公众普遍忽视。公园毗邻城市的足球场和体育中心，现已成为海尔兹利亚大型综合娱乐活动场所中不可或缺的一部分。

The park site is a historical flood basin draining toward the Mediterranean Sea via an ancient roman aqueduct (only last year a new tunnel was added). More than half of the surrounding town's storm drainage ends up in the park, channeled previously in concrete-lined runnels toward the aqueduct. Most of the former agricultural land has been abandoned or used as landfill for excess earthworks. Large winter ponds with standing water during the winter months exist on site due to the heavy clay soil of the area. The seasonal flooding has protected this piece of land from housing development in the center of an otherwise very densely populated area.

公园地处一个有着久远历史的排洪区，蓄滞的洪水经由这里，再通过一个古老的罗马时期修建的渡槽而流向地中海（直至去年，这个地区才增加了一个新的泄洪管道）。过去每当暴雨到来时，周边有超过半数的城镇，都会通过这个由混凝土衬砌渠道而形成河道排入公园，再流向渡槽。以前，大部分的农业用地都已经荒废了，这些土地被用作垃圾填埋场，堆放多余出来的土方。由于该地区的土质是那种十分　稠的　性土，其沁水率很低，一般大型的池塘可以在冬季的几个月时间里一直持有积水。正是这种季节性的洪涝灾害保护了这个中心区域，使其未被用作房地产开发。

南 "制革厂" 岛运动公园
Redesign Southern Tan Mill Island Sports Park

LOCATION:
Berlin, Germany
AREA:
6.5 ha
PHOTOGRAPHER:
Rehwaldt LA, Dresden
DESIGN :
Rehwaldt LA, Dresden

项目地点：
德国 柏林
面积：
6.5 公顷
摄影师：
Rehwaldt LA，Dresden
设计单位：
Rehwaldt LA，Dresden

Southern tan mill island (Lohmuehleninsel) is located in the Berlin quarter Kreuzberg. The island evolved from the construction of the state canal (Landwehrkanal) during 1845 and 1850. It is 600-meter long and 100-meter wide. The tan mills which were formerly located on the site gave the island its name. The flood channel (Flutgraben) which defines the western bank of the tan mill island used to be the Inner German borderline between East and West Berlin until 1989. After Germany was reunited, the former boarder strip was transformed into a park, which improved the quality of life in Kreuzberg for its lack of green spaces. Most reconstructions of the public open spaces on southern tan mill island were finished in the years of 2007 and 2008. The new design is a good addition to the existing areas for a greater variety of activities for all age-groups.

南 "制革厂" 岛（Lohmuehleninsel）位于柏林市克洛伊茨贝格区，从 1845 到 1850 年建造的国家运河（Landwehrkanal）演变而来，长 600 米，宽 100 米。这个岛屿以原来位于这一地区的 "制革厂" 命名。洪水河道（Flutgraben）界定了 "制革厂" 岛的西河岸，这里直到 1989 年还一直是东柏林和西柏林的德国内部边界。德国统一以后，由于这里缺乏绿地，原来的边界区改造成了一个花园，提升了克洛伊茨贝格区居民的生活质量。南 "制革厂" 岛公共空间的大部分改建工程都于 2007 年和 2008 年完成。新设计为原来的区域增加了色彩，为各个年龄段的人群提供了各种活动设施。

The southern tan mill island is a popular recreation area and a frequently used transit space for pedestrians and cyclists at the same time. For the subsequent functional conflicts a guideline for the whole southern tan mill island was developed.

南"制革厂"岛不仅是一个受欢迎的娱乐区，同时还是一个行人和骑脚踏车的人频繁使用的交通区。为了避免日后发生功能方面的冲突，相关人员为整个南"制革厂"岛制定了设计指导原则。

Considering the overall situation, new intensive uses on the island were to be limited. Every additional activity would have caused more usage frequencies and therefore further potential conflicts. So, it was important to place additional recreational offerings sparingly, sensitive and well arranged.

从全局考虑，新设计要对岛屿的集中使用进行限制。每个新增加的功能区都会频繁使用，并可能导致冲突。因此，要分散、合理地布置新增设的娱乐设施。

Additional planning targets were the improvement of paths connections, the functional redesign of entrance areas, the better connection with Goerlitz Park and the establishment of visual relations. For a better orientation and the ordinary safety shrubs and bushes were pruned. Additional plantings were placed only in selected areas.

其他规划目标包括改善小路之间的联系；对入口区域重新进行功能设计；与 GOERLITZ 公园之间进行更好的联系；建立起视觉联系。由于朝向和安全问题，对灌木丛进行了修剪。其他植物只能在选定的区域种植。

The wide main path as well as the playing strip were formed as generous active and transit spaces focusing on deceleration. The demands of different users are reflected in surface materials and features arranged on the playing strip. Cyclists, skaters and pedestrians share the spaces in slow motion or just pass by on the fast track in the northwest.

宽阔的主干道和条形娱乐区形成了一个宽敞的活动和交通空间，减速成为这里的主题。不同人群的需求反映在条形娱乐区的表面材料和特征上。骑脚踏车的人、滑冰的人和行人都可以在这里缓慢通行，或从西北侧的快速车道上通过。

Benches for resting are arranged along the main path. A large staircase partly equipped with wooden seats leads down to the channelside walk and can be used as a place for sojourn.

主干道的旁边设置了长椅，可以用来休息。一个巨大的台阶通向水渠边的人行道，上面设有木质的座椅，人们可以在这里驻足停留。

比尔梅公园
Bijlmerpark

ARCHITECT:
Carve
COLLABORATION:
Marie-Laure Hoedemaker
WORK TEAM:
Elger Blitz, Mark van der Eng, Jasper van der Schaaf, Lucas Beukers
LOCATION:
Amsterdam, The Netherlands
AREA:
8,400m²

建筑师：
卡韦（Carve）
合作人员：
玛丽—劳尔霍迪麦克
工作团队：
埃尔格·布利茨、马克·英格、雅斯佩尔·沙夫、卢卡斯·伯凯尔
项目地点：
荷兰阿姆斯特丹
面积：
8400 平方米

The Bijlmerpark is the main park in Amsterdam's southeast district "Bijlmermeer". This 1960s and 1970s modernistic suburb of Amsterdam, characterized by high-rise residential and disjunctive infrastructural networks for pedestrians, cyclists and motorists where services and facilities were few and far between, had developed numerous social problems by the end of the 1980s. Radical, integral restructuring process was initiated. The renewal of the Bijlmerpark is the final chapter in this process. a move from quantity to quality has become the policy for the redevelopment. Bijlmerpark was both to remain the main park in the built-up area and was identified as a new residential environment with a program of approximately 900 dwellings. The main components are a park encircling, a central sports facility and residential units along the flanks of the park. The concept reconfigures the spatial and social structure. The new housing is facing the park, providing eyes and ears on the park. The central position of the sports park keeps this facility within walking distance of the residents.

比尔梅公园是阿姆斯特丹东南部地区"拜尔美米尔"的主要公园。这个阿姆斯特丹的现代郊区在上世纪 60 年代和 70 年代以其高耸的住宅和行人、自行车与汽车相分隔的道路设施而闻名。这里服务和设备稀少，直至 20 世纪 80 年代末仍然存在许多社会问题，因此开展了大规模的重建工程。而比尔梅公园的修复则是这个阶段的最后一章。由数量向质量转变是再开发过程中的主要政策。比尔梅公园需要在建成区保留主园区，并且要建成一处带有 900 个住所的居住环境。其主要组成部分是围绕着核心体育设施的园区以及公园两侧的住宅单元，其理念是重新配置空间和社会结构。新的住宅区面向公园，居民可以听到并看到公园里的一切，走几步就能到达运动公园的中心位置。

The sport and game esplanade in the center of the Bijlmerpark is the hub for all playing activities. The sports and game esplanade is implemented as a bypass of the main route: the circular pedestrian and bicycle park-route. The esplanade embraces several elements: a ball court, the playing strip, the "King Crawler", a skatepark and a water and sand playground.

比尔梅公园中心的体育游乐场构成了所有活动的核心。体育游乐场也是公园主线的一条支路:具有圆形的人行道和自行车道。体育游乐场包含一系列元素:一个球场、游戏地带、"攀爬王"、滑板运动场以及水上和沙地游乐场。

The paved multi-sports court includes a stage and ball-catchers with "professional" grade boulder-routes. A series of yellow frames on bright and sparkling pink safety surfacing makes the playing strip, containing different types of rope bridges and a zipline connecting to the "King Crawler". the "King Crawler" structure is a multilevel playing wall that incorporates facilities for the playground manager and two public toilets. The playing strip is located at the foot of two rolling green hills with trees. On top of the hill a skatepark consisting of two connected pools is hidden, with banks and stairs coming down to ground-level again. On top of the other hill one can find a water and sand playground, a colourful landscape for the youngest children with sandboxes and waterjets.

铺平的多功能运动场包括一个舞台和带有"专业"鹅卵石小路的捕球装置。明亮而闪耀的粉红色安全平面上的黄色框架代表游戏地带,含有不同类型的绳桥和与"攀爬王"相连的滑索道。"攀爬王"是一个多层的游戏墙,包含游乐场管理设施以及两个公厕。游戏地带位于两座绿意盎然的小山的山脚处。其中一座小山的顶部有一个滑板运动场,由两个相连的水池构成,较为隐蔽,它的河岸跟阶梯与地面持平。在另一座小山的顶部,可以看到水上和沙地游乐场,犹如一幅五彩缤纷的风景画,最小的孩子带着他们的沙盒跟水枪在那边玩耍。

墨尔本儿童运动公园
Melbourne Sports Park

CLIENT:
City of Melbourne
SCOPE:
Design refinement, design development, contract documentation

委托方：
墨尔本政府

设计范围：
设计修改深化设计，合同文件

Nestled between the Melbourne Town Center Birrarung Marr and Federation Square developments, lies the new family-focused center and activity space, playfully titled ARTplay. The area was identified as a site for family and child-focused activity in the 1998 Birrarung Marr master plan, and was successfully opened to the public in November 2004.

项目场地位于墨尔本市中心 Birrarung Marr 公园和联邦广场之间，是一个供家庭进行娱乐活动的场所，因此命名为艺术游乐园 。根据 1998 年的 Birrarung Marr 公园的总体规划，现场的用地定为供家庭和儿童活动使用的区域，并在 2004 年 11 月成功向公众开放。

The project lies within an awkward triangular site, abutting a car park and at the foot of the monumental earth forms of the Birrarung Marr parkland. The project brief called for the creation of an imaginative landmark playspace to complement the ARTplay building and its programme of curated art workshops, and for it to be a place that the children of Melbourne would find memorable and exciting.

项目的现状是一个三角形的场地，毗邻一个停车场，在 Birrarung Marr 公园纪念雕塑下方。设计任务书要求将项目发展为一个可令孩子们轻松愉快的游乐场所，激发孩子们的想象力，并设有艺术游乐园相关的艺术工作室，给予墨尔本的孩子们难忘美好的回忆。

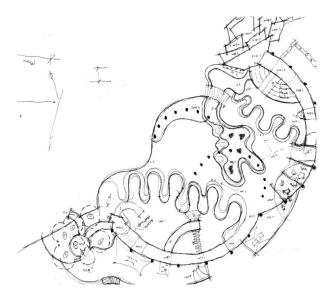

ARTplay is classified as having regional significance by City of Melbourne's Parks and Recreation division. Through extensive public consultation with the Playground and Recreation Association of Victoria (PRAV), stakeholders, families and children through workshops and a "play day" an extensive brief was developed outlining clear objectives that detailed the types of play and experiences required. It was recognised that imaginative play was just as important as active play, and that the value of a vibrant and stimulating environment was just as critical to the play experience.

艺术游乐园所在的位置是墨尔本的公园与休闲娱乐用地区域，意义重大。设计师与维多利亚休闲娱乐协会、相关机构、家长和孩子们进行了详细的公众咨询与探讨，共同完善了本项目的设计任务书《活动日》，确定了明确的目标以及详细的活动内容。大家一致认为能激发孩子们想象力的活动与体育运动同样重要，而且一个充满活力和想象力的环境与娱乐活动同样重要。

SECTION A-A

Scale 1:200

A-A 部分
比例 1:200

Timber boardwalk/ ramp
structure with associated
play equipment

木板路 / 斜坡结构
相连的游戏设备

Grass
mound
草堆

Large steel and timber
poles with the ability
to support play equipment
支撑树木的钢架

Existing building to
be renovated
现有建筑改造

The Artplay playground is situated between Federation Square and Birrarung Marr Park on the Yarra River, Melbourne. The aim was to provide a playground that could be creatively engaged with by children attending workshops at the associated Artplay building.

艺术游乐园位于联邦广场和墨尔本雅拉河畔的 Birrarung Marr 公园之间，其目的是要为儿童提供艺术游乐场所，激发孩子们的想象力，并让孩子们在艺术工作室中游戏，寓教于乐。

This is a unique playground that is not a collection of catalogue items, but rather a cohesive landscape approach involving carefully chosen materials, textures, colours, levels and spaces. Earthy tones, flowing forms and native vegetation support an Australian landscape theme that encourages interaction and experimentation with outdoor spaces.

这是一个独特的游乐园，在游乐园内并不只是设置一系列的游乐项目，而是通过综合的景观设计手法，精心选择材料、质地、颜色、层次和空间，从而完成游乐园的设计。通过大手笔地采用泥土色，流畅的风格和本土植被来体现澳洲景观的主题，鼓励人们多到户外活动。

The Artplay playground responds to the intensity of its location and borrows from the surrounding spaces, yet stands proud as an exciting, colourful and adventurous landscape.

艺术游乐园的景观设计大力发挥其地段的优势，综合周围空间的优势，发展成色彩缤纷的游乐园，带给孩子们一种愉快的、惊喜的体验。

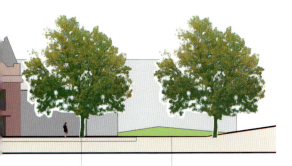

**Existing
Elm trees**
现有的榆树

Proposed turf mound
计划的木炭堆

尼古拉斯贝茨运动广场
Nicolaas Beetsplein Sports Grounds

LOCATION:
Oud Krispijn Dordrecht，The Netherlands
ARCHITECT:
NL Architects and DS Landschapsarchitecten
NL Architects: Pieter Bannenberg, Walter van Dijk,
Kamiel Klaasse
DS Landschapsarchitecten: Jana Crepon
TEAM NL:
Kirsten Huesig with Beatriz Ruiz and Maurice
Martell
TOTAL AREA:
9,400 m²

项目地点：
荷兰多德雷赫特市 Oud Krispijn

建筑师：
NL Architects and DS Landschapsarchitecten

NL Architects：Pieter Bannenberg，Walter van
Dijk，Kamiel Klaasse

DS Landschapsarchitecten：Jana Crepon

NL 团队：
Kirsten Huesig with Beatriz Ruiz and Maurice
Martell

总面积：
9400 平方米

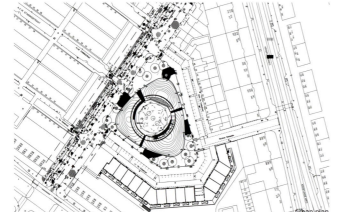

urban plan

Oud Krispijn is a neighbourhood in the city of Dordrecht. It was planned by Van der Peck in 1932 as a Garden City for the working class: a warped grid, small housing blocks, sweet little houses with front and back gardens, refined "corner solutions", alleys and several "green" axis and some modest squares. And a lot of hedges.

乌德克里斯金街区位于多德雷赫特市。1932 年由设计师 VAN DER PECK 设计，旨在为上班族打造一个园林街区：弯曲的铁网，小型住宅街区，静谧别致、前后都有花园的小房子，精致的街角小路，绿荫构成的轴线，合宜的广场，还有很多树篱。

In the eighties the district deteriorated and became one of the more problematic areas in the city, notorious for drug use and regular shootings. Here and there derelict original housing was replaced by "stone cold" architecture, partly destroying the initial charm of the area.

20 世纪 80 年代，乌德克里斯金街区每况愈下，变成了城市里一个问题频发的地方，毒品泛滥，枪击案屡发。原来处处可见的建筑也被冷落，取而代之的是冰冷的岩石结构，街区原有的魅力丧失大半。

People from numerous ethnic backgrounds found a place to live here, creating a vibrant Corner Shop culture: quite exotic. Once we wanted to buy a Coca Cola in one of these shops, but the owner at first didn't know what we were talking about: coke light, what could that be? We had to point it out in his fridge.

来自不同民族的居民聚居于此，形成了一个生机勃勃的街角商店文化：极具异国情调。有一次，我们到一个街角商店买可口可乐，但是店主没明白我们的意思，问："可乐光，那是什么东西？"后来我们不得不指着冰箱里的可乐，让他明白。

Interestingly enough many front doors are left open, habitually. An inviting gesture that could be part of the inheritance of the original Dutch population that likes to go visit the neighbours for coffee. This culture seems to be embraced by newcomers, creating an anachronistic, sociable atmosphere that you would never associate with the rough public image of this neighbourhood.

非常有趣的是，按照当地风俗每家每户的前门都是大开着的。土著荷兰人有一个沿袭已久的习惯，他们喜欢到邻居家拜访，喝杯咖啡。外来住户也很喜欢这一文化习俗，街区洋溢着与时代不符的社交气氛，让人很难将这种友好气氛与街区糟糕的公众形象联系到一起。

Some time ago, the planning department invited Atelier PRO and DS Landscape Architects to help the reconstruction of the emblematic arrangement of this garden city. One of the concerns was that the total amount of public space and playgrounds was not sufficient. It was suggested to remove one entire building block: the space for the Nicolaas Beetsplein was created. The local government invited DS Landschapsarchitecten to develop the design.

一段时间以前，规划部邀请 ATELIER PRO 和 DS 景观设计师为这座园林街区的重建献计献策。一个关键问题是公共空间和广场的总面积数不足。建议取消一整作住宅楼，于是 NICOLAAS BEETSPLEIN 应运而生。当地政府邀请 DS 景观设计师来做设计。

In this framework the CBK, the Art Center of Dordrecht initiated an invited competition to produce ideas for seven of the neighbourhood squares. NL Architects' proposal won; Beetsplein was going to be the first in the row. Gerrit Willems, the director, anticipates a new role for Art in Public Space: influencing the "program" and the usage seems more to the point then just affecting the appearance. The CBK paid the design fee of NL Architects and contributed substantially to the construction costs. As such the collaboration between the two offices was conceived.

在此框架中，多德雷赫特艺术中心 CBK 邀请诸多设计室为它的七个广场提供设计。NL 设计师的提案得到了认可，BEETSPLEIN 是首个设计项目。总负责人 GERRIT WILLEMS 先生期望"公共空间的艺术"能扮演新的角色，不但改变城区的外观，也能为项目结构和它的功能增色。CBK 支付了 NL 设计师的设计费，并承担了大部分建筑费用。两家公司的合作方式也确定了下来。

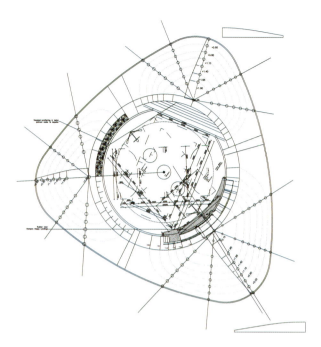

德国联邦园林公园
BUGA

LANDSCAPE:
ARCHITECT ATELIER DREISEITL
TEAM MEMBERS:
**GERHARD HAUBER, HENDRIK PORST,
CHRISTOPH HALD, STEFAN BRUCKMANN,
GUSTAVO GLAESER, CHRISTOPH WURTLE,
ANGELIKA BUCHELE, SEBASTIAN ZIEGELMEIER**
LOCATION:
KOBLENZ, GERMANY
AREA:
1,111m²
PHOTOGRAPHER:
ATELIER DREISEITL, THOMAS FREY, DOHERTY

景观设计师：
德国戴水道设计公司
团队成员：
格哈德·哈芭、昂德里克·普斯特、克里斯托夫·哈尔德、斯蒂芬·布鲁克曼、古斯塔沃·格莱泽、克里斯托夫·沃尔托、安格莉卡·普谢礼、塞巴斯蒂·安赞格米尔
项目地点：
德国科布伦茨
景观面积：
1111m²
摄影师：
德国戴水道设计公司、托马斯·弗雷、多尔蒂

Water is ubiquitous at Deutsches Eck in Koblenz where the two rivers, the Rhein and the Mosel, meet and join to one. We are surrounded by water when standing at this impressive and historic landmark. But is it possible to physically interact with the water here? Through the expanded and renovated riverside promenade, the rivers are close, but in order to live up to the motto of the Bundesgartenschau "Koblenz Transforms", the water has to become accessible, touchable and tangible. In our view, this experience should be realized with the water playground which is situated only a few steps away from the Deutsche Eck.

科布伦茨的德意志之角到处都是水，莱茵河和摩泽尔河两大流域在这汇合。站在这片令人印象深刻的、具有历史意义的土地上，我们可以看到周围全是水。然而我们能用身体与这里的水进行交流吗，通过这条扩张和翻新过的步道，我们离水更近了，但是要达到德国联邦园艺博览会的座右铭"科布伦茨的革新"，我们必须将水变得有形而可触碰。我们认为这种体验可以通过位于德意志之角不远处的水上乐园来达成。

© Tomas Frey

The starting point of our work was to develop a place to experience water, incorporating the theme of the two rivers, the Rhein and the Mosel. It was thought to present different landscape and water characteristics of the rivers and hereby work on the differences between the two rivers. In addition, beside the permanent construction of this water playground, a temporary exhibition ship would display the work of the water federation and ship administration to visitors.

作品的出发点是开发一处能体验水的地方，且融入莱茵河和摩泽尔河两大流域的主题。这个地方必须要能体现河流不同的景观和特征，继而呈现出两条河之间的不同。另外，水上乐园这座永久性建筑的旁边还有一艘临时展览船，向游客展示了水联合会和船舶管理处的作品。

梅利斯斯托克运动公园
Playground Melis Stokepark

LOCATION:
Hague The Netherlands
AREA:
810 m²
DESIGNER/ARCHITECT:
Carve took care of design, engineering and site
management and surveying
TEAM:
Elger Blitz, Mark van der Eng, Renet Korthals
Altes, Jasper van der Schaaf, Amber van Capelle
PHOTOGRAPHER:
Carve

项目地点：
荷兰海牙

面积：
810 平方米

设计师 / 建筑师：
Carve took care of design, engineering and site
management and surveying

团队：
Elger Blitz, Mark van der Eng, Renet Korthals
Altes, Jasper van der Schaaf, Amber van Capelle

摄影师：
Carve

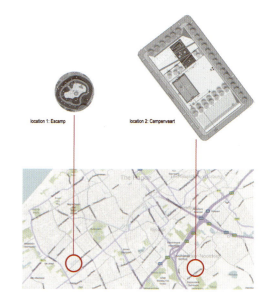

Carve was asked by the municipality of the Hague to design two "integrated play facilities", playgrounds suitable for children with and without disabilities. How does one design a playground where the difference in play between children with and without disabilities is eliminated? A playground that offers challenges and appeals to children with and without disabilities? In Carve's vision "playing together" doesn't mean playing next to each other.

Carve 应海牙市政府之邀，为残疾儿童和非残疾儿童设计两个配有"综合性娱乐设施"的运动场。如何让所设计的运动场上毫无残疾儿童和非残疾儿童的差别呢？如何保证运动场对所有的儿童都充满吸引力和挑战呢？Carve 最终的答案是"一起玩"，这并不是指你挨着我，我挨着你的一起玩。

Both playgrounds are designed with the capacities of children with a disability in mind, and not their limitations. Every child wants to explore its possibilities and wants to be able to extend its boundaries. Importantly, the playgrounds do not directly show that possible limitations (visual, auditory, physical and mental) have been taken into account. They offer increasing challenges for all children and thereby contribute to play without limitations. With a completely authentic language of form and choice of colour they stimulate curiosity and allow children to discover the extensive playing facilities by and with each other.

在两个体育场的设计上，我们自始至终考虑的是残疾儿童的不便，而不是给他们的能力设限，因为每个孩子都渴望发现自己的可能性，想挣脱自身的束缚。重要的一点是，运动场上不应该直接体现出在视觉、听觉、身体和精神上的局限性。它们要为所有儿童制造挑战，让孩子们尽情玩耍。出奇的形式、合适的颜色相辅成，一起激发儿童的好奇心，让他们共同探索游乐设备的奥妙。

The playground consists of an ascending ring that is both a curving route to the slide, as a two-sided climbable boundary. The vertical outer wall is made of strips of wood with perforations and round climbing holds. The ring encloses an inner area with blue undulating playing slopes and a sandpit. Several passages lead to and from this inner area and offer their own seating and playing possibilities. The privacy of this inner play area offers a secluded spot for children who have difficulties keeping up in the big open spaces.

直墙体是木条构成的，上面有小孔和圆的攀岩扶手。平台内部设计了蓝色的起伏不定的斜坡和一个小沙坑。几条线路连接内部区域和外部空间，线路两侧还设有座椅和游戏设施。有些儿童不喜欢在大庭广众下玩，那么荫蔽的内部空间就是个很好的选择了。

The wooden wall with horizontal climbing routes and steep slopes create playing possibilities that each child, depending on his own skills, can use in different ways. Inside, around and on the ring is wide space for challenging active games, repetitive movement games (turning, sliding, jumping and swinging), construction play (sand) and for fantasy games (tunnels, platforms, shelters).

一面木墙，两种攀登方式。孩子们可以使出浑身解数，用不同的方法，沿着木墙在水平方向上移动，或沿着陡坡攀登。平台上及四周留有足够的空间，孩子们可以做活跃的游戏，重复类的运动（比如转圈、滑行、跳跃和荡秋千），还可以用沙子建城堡，或者到隧道、舞台和其他荫蔽的地方体验刺激。

Pimelea 运动场西悉尼公园
Pimelea Play Grounds Western Sydney Parklands

LOCATION:
Sydney Australia
AREA:
2.00 ha
DESIGN COMPANY:
McGregor Coxall

项目地点:
澳大利亚悉尼
面积:
2.00 公顷
设计公司:
McGregor Coxall

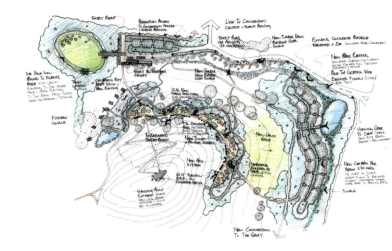

The original Pimelea parklands was constructed prior to the 2000 Sydney Olympics. The Western Sydney Parklands Trust engaged McGregor Coxall to revitalize and extend the parks facilities by first undertaking a master plan and then designing and documenting a portion of the master plan. This included an upgrade and extension to the existing toilet block, BBQ and picnic facilities, along with an extensive redevelopment of the children's play area. The works also developed a design for a new toilet block and shade elements. A sensitive response to the rural nature of the site, underpinned by a strong sustainable strategy, drove the site's redesign. Power for the site is generated by solar panels, toilet flushing utilises dam water, all grey water is reused for irrigation and recycled materials are used where possible. The play area extends this concept by introducing recycled water through a children's play pump and water course system within a unique play experience that both excites and delights.

原有的 Pimelea 游乐场建于 2000 年悉尼奥运会之前。悉尼公园联合组织委托 McGregor Coxall 建筑事务所负责让公园重新焕发生机并完善公园设施。McGregor Coxall 需先制定蓝图，然后设计并确定蓝图中部分地区的设计图纸。这部分包括提高现有公共卫生设施条件，扩大其面积；完善野外烧烤和野餐设施以及扩大儿童游戏区。设计师还提出了新的公共卫生间和林荫设计，以强烈的可持续战略为基础，灵活地呼应项目地点的乡村风光是设计师所做设计的主旨。场地采用太阳能电池板发电，卫生间使用大坝水冲洗，中水再利用用于灌溉，所有可以循环利用的材料都尽其所能。环保的理念在游戏区得到进一步利用，儿童做游戏的水泵和水道里面引入的都是循环用水，既可让孩子们畅快玩耍，也于环保有益，一举两得。

皇家山公园游乐场
Mount-Royal Park's Playground

LANDSCAPE ARCHITECT:
CHA CARDINAL HARDY
LOCATION:
MONTREAL, CANADA
CLIENT:
CITY OF MONTREAL
PHOTOGRAPHER:
MARC CRAMER

景观设计师：
CHA CARDINAL HARDY

项目地点：
加拿大蒙特利尔

委托方：
蒙特利尔市

摄影师：
MARC CRAMER

The theme is the Blue Spotted Sala mander, an amphibian native to Mount Royal and the starring feature which organises the play structures and other park elements. Water features and other innovative play structures are integrated into the silhouette of the salamander as it rises from the earth; this instigates a different kind of play, which encourages the children's motor, cognitive and social development.

设计主题为蓝色斑点火蜥蜴，这是生活在皇家山的一种两栖类动物，也是构成游乐场建筑和其他公园元素的主要特征。从地表升起的火蜥蜴轮廓中融入了水景和其他创新的游乐场结构；这是一种与众不同的新玩法，可以促进儿童的动机、认知和社会发展能力。

Beyond simply contending with a heritage site, the project highlights the therapeutic influence of this large scale green space in the city. The design was based on two distinct projections of the space; a vertically nuanced integration into the surrounding environment, and a horizontal plain contrasting the natural surroundings with the silhouette of the salamander and its bright colours. The neutral and discrete colouring of the structures allows them to melt when superimposed on the Olmsteadian decor.

除了对付这片历史地段外，项目还强调了这片大型绿地对于城市的治愈作用。设计基于空间内两个独特的地方：其中一个微微垂直地嵌入周围环境中，而另一个则是水平方向的平地，与火蜥蜴的轮廓及其鲜艳的颜色形成对比。

皇家山公园游乐场
Mount-Royal Park's Playground

The built area was as limited as possible in order to keep the ecological footprint to a minimum and the use of natural materials on the ground let surface water percolate into the soil. At the edge of the clearing, an ecological corridor linking two main nodes of a local ecological network was planted with indigenous species in an effort to regenerate the understory. Against this unusual backdrop, the landscape architect designed a Children Rights promenade of didactic elements.

建筑的颜色呈中性，较分散，可以很好地融入奥姆斯特德布局。项目尽量控制建筑区域，以减少生态足迹和对自然资源的使用，且保证地表水渗入土壤中。在空地的边缘有一条生态廊，连接当地生态网络的两个主要节点，种有各种本土植物，使下层植被得以再生。在这个特殊的背景下，景观设计师设计了一条儿童权利步道，充满各种说教的元素。

Public interpretive panels allow people of all ages to discover the rights guaranteed to children by the International Convention of Children's Rights.

通过公共讲解板块，各个年龄段的人群都可以阅读《国际儿童权利公约》颁布的儿童保障权利。

Coordinated by the landscape architect, three posters, two brochures, a questionnaire and a power point presentation were developed for public meetings.

在景观设计师的协调下，设计人员为公共会议开发了三张海报、两本小册子、一份问卷调查以及一个 PPT 演示。

Almost 200 people attended the information meetings and consultation activities. The participants endorsed the proposed design for this project. Built over a period of 16 months, this project recognises the emblematic and symbolic value of "The Mountain". Both a school group and elected official were part of the opening ceremonies of the project; an event which promoted both the profession of landscape architecture and the rights of children.

将近 200 人参加了信息发布会和咨询活动，参与者均支持项目的提案。经过 16 个月的努力，项目呈现了"山峰"的象征价值。学校群体和当选的官员都参加了项目的开幕仪式，活动不仅提升了风景园林的专业程度，而且还宣传了儿童的权利。

Vlaskamp 运动公园
Vlaskamp Sports Park

LOCATION:
Hague The Netherlands
PARK AREA :
30 hectare
DESIGNERS:
Carve

项目地点：
荷兰 海牙

公园面积：
30 公顷

设计公司：
CARVE

water
bushes
ground cover
existing trees
grass
pathways
boulders
benches
play islands with safet

斯塔万格市地质公园
Stavanger Geological Park

LOCATION:
Norway
AREA:
2,500 m^2

项目地点：
荷兰海牙

面积：
2500 平方米

The Geopark directly applies three different types of resource. First, the industry's geological and seismic expertise, second, it's production and handling of technology, materials and waste related to offshore-platforms, and third, the ideas of several youth groups and young individuals for a future park in the city center.

地质公园直接采用了以下三种不同的建议：第一，了解有关地质和地震方面的专业知识；第二，掌握石油业技术、石油生产知识和近海钻井平台附近废弃物等方面的知识；第三，一些青年团体和年轻人表达了这样的看法——他们希望将其建成一座城市中央公园。

An interaction between these resource groups and Helen & Hard resulted in a 2500 sq.m. waterfront youth-park and outdoor science center for the adjacent Norwegian Petroleum Museum.

通过与提出上述建议的"智囊团"的沟通，最终 Helen & Hard 提出建造一个占地 2500 平方米的海滨公园方案。它既是年轻人的公园，也是相邻的挪威石油博物馆的户外科学中心。

An initial intention was to give a tangible experience of the oil and gas reservoir "Troll", hidden 2000–3000 metres below the seabed. The geological strata and associated drilling and production technology, reconstructed in a scale of 1:500, gives the outlines of the primary topography for Geopark.

地质公园的首要意图是让市民能亲身体验石油天然气的开采。特罗尔油田位于海底 2000~3000 米处，是北海最大的油气开采基地。设计师煞费苦心地按照 1:500 比例设计了一个特罗尔的模型，展现了地层结构、钻井技术和生产技术。特罗尔模型勾勒出了地质公园的基本地势轮廓。

This "geo-landscape" is further developed in a sequence of playful and empirical steps, and programmed in workshops with youth groups for various activities like biking, climbing, exhibition, concerts, jumping, ball play and "chill-out" areas. The oil layer in the Troll field, with its drilling wells, is represented as a skating park, and the geological folds are reused as exhibition walls for graffiti and street art. The surfaces and installations are reconstructed out of recycled and reshaped elements from the petroleum installations, the abandoned Frigg platform, offshore bases, equipment suppliers and scrap heaps.

设计师运用一系列游戏和实验性措施，进一步提升了公园的"地质景观"特点。通过和青年团体的研讨，设计师决定让公园具备更多活动功能，包括骑车、攀爬、展览、举办音乐会、跳跃、开展球类运动和休憩。特罗尔油田模型及钻井可作为溜冰场，"地质褶皱"可作为涂鸦和街头艺术的展览墙。对石油设备、废弃的钻井平台、近海油气基地和废弃物进行回收再利用和加工改造，使其在公园中实现新的用途。

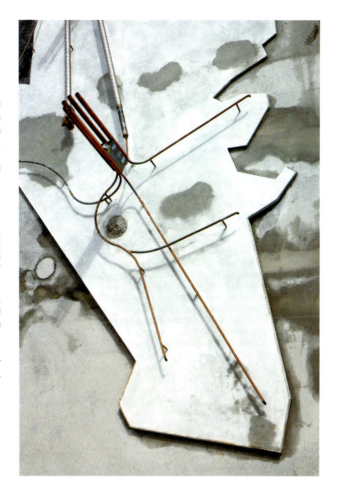

The park is thriving. Kids, parents and youngsters are using the park at all hours, turning a formerly abandoned site into a humming social meeting point. Local newspapers, politicians and park users are now fighting to make the park, which was originally planned as temporary, a permanent feature.

地质公园生机勃发。这里每时每刻都能看到孩子、家长和年轻人，一个昔日被废弃的地方而今成了居民的聚集点。公园原本是一个临时性建筑，现在当地的报刊、政客和逛公园的人都在为让它成为永久性景点而努力。

宫殿运动公园
Palace Sports Park

LOCATION:
Hague，The Netherlands
AREA:
250 m^2
DESIGNER/ARCHITECT:
Carve took care of design, engineering and site management and surveying
TEAM:
Elger Blitz, Mark van der Eng, Renet Korthals Altes, Jasper van der Schaaf, Milan van der Storm, Thomas Tiel Groenestege
PHOTOGRAPHER:
Carve

项目地点：
荷兰 海牙
面积：
250 平方米
设计师 / 建筑师：
Carve took care of design，engineering and site management and surveying
团队：
Elger Blitz，Mark van der Eng，Renet Korthals Altes，Jasper van der Schaaf，Milan van der Storm，Thomas Tiel Groenestege
摄影师：
Carve

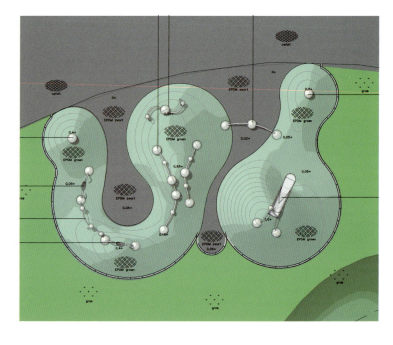

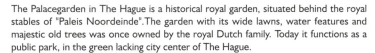

The Palacegarden in The Hague is a historical royal garden, situated behind the royal stables of "Paleis Noordeinde".The garden with its wide lawns, water features and majestic old trees was once owned by the royal Dutch family. Today it functions as a public park, in the green lacking city center of The Hague.

海牙的宫殿花园是一座具有历史意义的皇家园林，位于皇家马房"Paleis Noordeinde"身后。花园里草地疏朗，几处流水设计，古树参天，曾一度是荷兰皇家的御花园。如今的宫殿花园已是一座面向公众开放的花园，位于绿色匮乏的海牙城中心。

Carve was asked by the city council to design a playground especially for the youngest of age (children between 1-6 years old). A sculptural character and the preservation of the authenticity of the garden were specific requirements for this design to meet.

Carve 建筑设计工作室受海牙市委员会之托为宫殿花园设计一个操场，目标人群主要是 1 至 6 岁的儿童。既要保留公园的真实性，也要具有雕塑特点是本次操场设计需要满足的两项要求。

The design involves a simple play structure of several white spheres, which are connected by curved tubes. Swinging, sliding, rocking. Together with the sloping safety surface the pearls form an adventurous play area, providing various play functions and an atmosphere where children might find their imagination stimulated.

设计包含一个简单的游戏设施——几个用弯曲的管道连接的白色球体。孩子们可以在这里荡秋千、滑滑梯或者坐旋转木马。球形建筑犹如乳白色的珍珠，其倾斜度在安全范围内。这是一个很刺激的游戏区，不但具备多种游戏功能，还能激发孩子们的想象力。

Meant as an appeal to the sense of touch, the different materials give a different tactile sensation. The stainless steel tubes change temperature easily and have a hard, shiny polished surface, while the rubber coating of the pearls feels warm and has a softer surface.

为了增强孩子们对触觉的感知，我们采用不同的两种材料来产生不同的触感。不锈钢的管道温度变化快，表面坚硬光滑、闪闪发光；"珍珠"外面裹着是一层橡胶，表面柔和，摸起来温暖。

Placed in a corner of the park, the structure keeps clear of the broad lawns of the old garden.

儿童操场位于公园一角，避开了这座古老公园中宽阔的草坪。

威尔明顿海滨运动公园
Wilmington Waterfront Sports Park

LOCATION:
Los Angeles，CA
AREA:
30 acre
PROJECT TEAM:

Steve Hamwey	Nancy Fleming
Owen Lang	Tim Stevens
Vitas Viskanta	Melissa McCann
Caleb Bruner	Mark Eischeid
Raphael Justewicz	Joon Yon Kim
Chang Keun Lee	Conard Lindgren
Meghen Quinn	Simon Raine
Nitza Thien	Grace Leung
Tomer Maymon	Scott Odom
Angel Cantu	Zach Chrisco
Chuck Coronis	Michelle Gauvin
Oswaldo Palencia	Jose Miranda

项目地点：
加利福尼亚州洛杉矶市

面积：
30 英亩

项目团队：

Steve Hamwey	Nancy Fleming
Owen Lang	Tim Stevens
Vitas Viskanta	Melissa McCann
Caleb Bruner	Mark Eischeid
Raphael Justewicz	Joon Yon Kim
Chang Keun Lee	Conard Lindgren
Meghen Quinn	Simon Raine
Nitza Thien	Grace Leung
Tomer Maymon	Scott Odom
Angel Cantu	Zach Chrisco
Chuck Coronis	Michelle Gauvin
Oswaldo Palencia	Jose Miranda

Once a part of the Pacific coastline, the Wilmington community became disconnected from the waterfront by the Port of Los Angeles – a burgeoning, diverse mix of industrial maritime facilities. After completing the Wilmington Waterfront Master Plan, Sasaki identified three open spaces for implementation: the Wilmington Waterfront Park, the Avalon North Streetscape, and the Avalon South Waterfront Park. The Wilmington Waterfront Park is the first project to be fully implemented. Built on a 30 – acre brownfield site, the new urban park revitalizes the community and visually reconnects it to the waterfront. The park integrates a variety of active and passive uses – informal play, public gathering, community events, picnicking, sitting, strolling, and observation– determined through an extensive community outreach process. The open space serves as a public amenity by doubling the current community open space while also buffering the Wilmington community from the extensive Port operations to the south.

威尔明顿社区曾是太平洋海岸线的一部分，后来，日益繁盛的洛杉矶港将它与海滨隔离开来。洛杉矶港是一个项目多种多样的沿海工业区。在完成威尔明顿海滨公园的整体规划后，Sasaki 设计了三个有待实施的公共开放项目：威尔明顿海滨公园、阿瓦隆北区街景和阿瓦隆南区海滨公园。威尔明顿海滨公园是第一个全面实施的项目。建园用地是一块 30 英亩的废弃用地，这座崭新的公园将让整个社区焕发活力，将公园和海滨在景观上结合起来。无论你喜静喜动，公园里都有适合你的去处。这里可以举行非正式表演、聚会、社区活动、野餐，也可以坐着、漫步甚至看看风景。在进行广泛的社区调查之后，设计师为公园确定了广泛的功能范围。这个开放的公共空间将社区现有的公共用地面积扩大了一倍，它既是一个不可多得的公共空间，也是威尔明顿社区里从洛杉矶港到南部的缓冲地带。

Wilmington Waterfront Park

In order to protect the community park from the Port's impacts, Sasaki created a strong sculptural landform which elevates the existing planar grade of the neighbourhood to 4.9m. This land integrates a series of multipurpose playfields with shade-dappled, gentle grass slopes. Atop the landform, the El Paseo Promenade provides a primary component of the pedestrian circuit with seating, display gardens, and a shared use pedestrian and bicycle path linked to the California Coastal Trail. Tree-lined promenades extend the park's network of pedestrian circuits and meanders, offering a variety of seating for respite, contemplation, and viewing park activities including interactive water features, an adventure playground for children, plazas for gathering and performances, and picnicking within the tree groves. Datum Walk provides a central pedestrian axis traversing the park and connecting two park pavilions. The pavilions frame outdoor rooms that offer a variety of informal seating, shade, a dry concession, public restrooms, and three flexible, formal performance venues.

为了保护公园免受洛杉矶港的影响，Sasaki 创造了一个雕塑式地形，将附近原有的地面提高到 4.9 米高。这块场地里有各类运动场，运动场中点缀着绿树，长草的斜坡起起伏伏。在场地上面，是 EL Paseo 步行道，它是园中环形步行道的重要组成部分。在步行道两旁，散布着椅子、展览花园和一个自行车与行人混合用道——该道与加利福尼亚海滨道路相连。翠树夹道的步行道扩展了公园的环形道路网。人们可以漫步，也可以在座椅上休息，想想心事，看看周围的活动——这里有一个互动式水景观，一个儿童冒险乐园，几个集会和表演的广场，树林间的空地也常有游人野餐。基准大街横穿公园，连接两个凉亭，是公园的中轴线。公园的户外建筑结构以凉亭为主，有的凉亭可以让人们休息、为人们遮阳，有的亭子是公共卫生间，还有三个亭子是承办正式演出的地点。

Sasaki integrated sustainable design practices and innovative engineering technologies into the overall project. Stormwater management directs water to primary landscape zones to promote infiltration rather than municipal treatment, demolished paving was ground and reused for paving sub-base, and all plant materials were selected as ecologically adapted, indigenous, or salt tolerant and irrigated by reclaimed water. Building and site lighting highlights key park elements, reducing energy demands and light pollution through high optical efficacy. Along the Port's industrial edge, colourful planes forming the terrace walls are coated with titanium oxide （TiO_2）, which transforms harmful air pollutants to inert organic compounds innovative photocatalytic technology.

Sasaki 将可持续设计项目与创新的工程技术结合于公园的整体设计之中。雨水管理体系将雨水导入主要的景观区，而不是让其流入市政污水处理系统。拆毁的路面被用来作新路的地基。植被也选择适应生态环境、耐盐性强且适应再生水灌溉的本土植物。照明系统也体现了公园的核心要素，提高自然光的功效来降低能耗和光污染。在洛杉矶港的工业带边缘，彩色的平面构成了阶梯，上面涂着二氧化钛，利用先进的光催化技术将有毒的空气污染物转化为惰性成分。

市政公园景观
MUNICIPAL PARK LANDSCAPE

阿姆斯特丹米尔公园
"Meer-Park" Amsterdam

LOCATION:
Amsterdam the Netherlands
AREA:
24,000 m²

项目地点:
荷兰 阿姆斯特丹
面积:
24000 平方米

It was a small array of interventions which lead to the successful transformation and upgrading of the area. Firstly, all fences were torn down, removing property claims. It was agreed with the different sport clubs involved that they would continue to maintain their respective fields and would continue to have privileged training rights. When no scheduled trainings take place however, literally everyone can come to make use of the fields – 24 / 7. This development can be seen as an example of political vision.

通过一系列的项目介入,设计师顺利地完成了此区域的转变和重建。首先,在无任何产权纠纷的前提下拆除所有围墙。市委员会与相关俱乐部达成协议,各俱乐部依旧对自己的运动场地保有所有权和优先训练权。但在无任何训练日程的情况下,所有人都能够进入运动场活动。这一举措极具政治眼光。

By relocating a bike path inside an existing alley of trees, Carve created an open stretch of green – the urban axis of 1930 was again visible, which was then furnished with oversized benches, tables and barbecue facilities. This was done to respond to a trend which is exemplary in the Netherlands: a decreasing amount of urban open space has to accommodate a growing population in a densifying city with a growing need of urban citizens to recreate in the green. At the same time parks are used more intensively, partly due to changing demographics, where green areas play an increasingly important role in everyday recreation.

Carve 在原有的树丛里重新开辟了一条自行车道,树林原有的绿色越发开阔,20 世纪30年代的城市中轴线也再次出现在视线里。设计师还沿中轴线安放了超大的长凳、桌子和烧烤设施。在荷兰,人口和公共空间的矛盾风头正劲:在人口密度极大的城市里,日益减少的城市公共空间却要成为越来越多的市民休闲放松的去处,其接待能力实为有限。与此同时,公共绿地在选择住宅时发挥的重要作用,使公园附近的人口密度极大,造成了公园的超负荷使用。

Modern urban parks have an ambiguous character, an interesting landscaping is required and a spacious layout demanded, but they must answer many urban functions. Meerpark had to take recreation, relaxation, meeting, education, alternative sports and playing facilities – all on a relatively small piece of park into account.

现代城市公园的定义不太明确，有的强调迷人的风景，有的则重视广阔的空间、开阔的布局，但无论如何，真正的城市公园必须提供多项城市功能。于米尔公园而言，娱乐、休闲、聚会、教育、运动和比赛设施等因素都应该在议题之中，自有其空间。

The large central hill addresses this duality by integrating both aspects: one half is a grass slope with natural play elements scattered around – integrated with the surrounding nature. The other half has an artificial 3.5 m high wall in bright orange designed for professional bouldering, but appealing to climbers of all levels and ages. The redevelopment of this central axis marks the beginning of a further transformation of this area.

园中央的一个小山展现出米尔公园的双面风格：小山一侧是一片倾斜的绿草，四周散布各类运动设施，与周围环境相得益彰；小山的另一侧是一面亮橙色、高 3.5 米的人造墙体，是一个经专业雕琢的巨砾，却吸引了不同年龄段、不同程度的游客来此攀岩。再建后的中轴线是公园里翻天覆地变化的一个缩影。

Carve designed a network of connections through the maze of sport fields. The most adventurous connection is surely the play facilities criss-crossing over the water adjacent to the playground. But to generally unlock the potential of the in between passages through the existing undergrowth and trees, the accessibility of the park as a whole had to be enlarged. A network of nature walks, educational pathways, running tracks and adventurous trails are indicated at the entrances, available online and marked throughout the whole area. Information about history, vegetation and animals that live here can be found on especially designed information boards. The routing system was planned to evoke people to become aware and discover.

在迷宫般分散的运动场上设计了一个贯通线路网。从横跨河上的游戏设施过渡到地面的路线非常刺激。如若借助现有的下层灌木和其他树木，充分利用道路之间的空间，那么必须保证穿梭于公园交通便捷。公园入口、互联网上和公园里面都明确地表示了公园里的各种路径：感受自然的步行小路、富于教育意义的观赏线路、跑道和充满冒险色彩的路线。公园还设有信息板，讲述公园的历史，介绍园中植被和栖息的动物。公园路网的设计旨在激发游人发现探索的意识。

Alai Txoko 公园
Alai Txoko Park

LOCATION:
Irun Spain
SITE AREA:
30,000 m²
DESIGNER:
Lur Paisajistak S.L
PHOTOGRAPHER:
Lur Paisajistak S.L
DESIGN COMPANY:
Paisajsitak LUR, SL

项目地点:
西班牙 伊伦

占地面积:
30000 平方米

设计师:
Lur Paisajistak S.L

摄影师:
LUR PAISAJISTAK S.L

设计公司:
Paisajsitak LUR，SL

The old garden had been colonized by natural vegetation, creating a closed landscape printed insecurity in residents. The park project part the need to create an open space to provide the desired sense of urban safety.

公园原先是一大片天然植被，这种封闭的景观给居民烙印下深深的不安全感。该公园项目旨在打造出开阔的空间，带来人们渴望已久的都市安全感。

It creates a large open space colonized by grass and planting grass and cherry trees that occupies the central park. In the central lawn is sorted hospitality towards areas that help to soften the slope through the walls that also serve as banks.

天然草地和栽种的草坪共同形成了此开阔空间，公园中心还栽植了樱桃树。中央草坪笑迎八方来客，它也让充当堤岸的高墙显得不那么陡了。

In the more limited park space is defined by building a playground. The park incorporates into its planning the route of a bike lane.

在较为有限的空间修建一个运动场。公园的整体设计上还包括一条自行车道。

Ballast Point 公园
Ballast Point Park

LOCATION:
Sydney Australia
AREA:
2.8 ha
TEAM:
Philip Coxall, Adrian McGregor, Christian Borchert, Jeremy Gill, Kristin Spradbrow
DESIGN COMPANY:
McGregor Coxall

项目地点：
澳大利亚悉尼
面积：
2.8 公顷
团队：
Philip Coxall，Adrian McGregor，Christian Borchert，Jeremy Gill，Kristin Spradbrow
设计公司：
McGregor Coxall

Ballast Point Park is a stunning new harbourside destination, delivered by Sydney Harbour Foreshore Authority on behalf of the NSW Government and designed by McGregor Coxall. The park is a result of community action that stopped development of the site for residential development and returned the land to people of Sydney as a parkland. Ballast Point Park was opened to community acclaim in July 2009. This project involved McGregor Coxall leading a multidiscipline team in developing the design for this 2.8 ha new public park on the former site of a Caltex oil storage and grease manufacturing plant on Sydney Harbour.

Ballast Point 公园是让人叹为观止的海滨胜地。该公园由悉尼海港局代新南威尔士州政府所建，由 **McGregor Coxall** 设计。先前曾发生一起社会运动，要求将土地归还给悉尼人民，作为公共的公园用地，停止住宅区的扩张。**Ballast Point** 公园正是这场运动的结晶，它于 **2009** 年 **7** 月正式对游人开放。**McGregor Coxall** 带领一个多领域团队，共同完成这个占地 **2.8** 公顷的公园设计。该公园选址在悉尼海港原德士古储油仓库和油脂生产工厂的旧址上。

The design uses world leading sustainability principles to minimize the project's carbon footprint and ecologically rehabilitate the site. The design reconciles the layers of history with forward looking new technologies to create a regionally significant urban park. The environmental approach is further underpinned by site-wide stormwater biofiltration, recycled materials, and wind turbines designed for on-site energy production.

本设计遵循世界先进的可持续原则，将项目的碳排放量降到最低，恢复该地区的生态状况。本设计融历史特点与前瞻的新技术于一体，力求创造一个非凡的区域城市公园。公园四周的雨水过滤措施、可循环材料的使用及风力涡轮机发电等方式将进一步巩固公园的环境措施。

The design challenges our perception of materials and their use. Dominant new terrace walls sit atop the sandstone cliffs but these walls are not made of precious sandstone excavated from another site, rather from the rubble of our past. What once was called rubbish is now called beautiful. It is the new ballast. But it is more than this at play: It is the total composition of these recycled rubber filled cages, off set with concrete coping panels topped with fine grain railing, that allow these walls to sit confidently at the portal to the inner harbour.

本设计冲击了我们对原料及其应用方式的原有概念。引人注意的砌在砂岩峭壁上的新式扶梯挡土墙并不是从别的地方挖掘、运往这里的，而是一些碎石。正是变废为宝的方式让它们重新履行了铺路石的使命。但是它们的意义并未止步于此：循环使用的碎石填满笼子，然后注入水泥，填满缝隙，在碎石板的顶部铺上细沙，这些步骤的共同作用让这些墙体牢牢地屹立在通往内部港口的大门周围。

8 vertical axis wind turbines and an extract from a Les Murray poem, carved into recycled tank panels, forms a sculptural re-interpretation of the site's former largest storage tank. The wind turbines symbolise the future, a step away from our fossil fuelled past towards more sustainable renewable energy forms.

8 个纵轴风力涡轮机，可循环的再生槽板，这些槽板上刻有莱斯·穆瑞诗中的一个句子，共同对这个场地原有的大型储油罐做出了雕塑性的再阐述。风力涡轮机象征着未来，一条逐渐远离原有的石油燃料、趋于可持续能源的新道路。

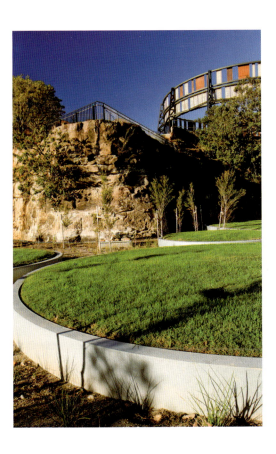

魁北克公园
Promenade Samuel-de Champlain

ARCHITECT:
Consortium Daoust Lestage + Williams Asselin Ackaoui + Option Amenagement
LOCATION:
Quebec, Canada
PHOTOGRAPHER:
Marc Cramer

建筑设计：
Consortium Daoust Lestage + WilliamsAsselin Ackaoui + Option Amenagement

项目地点：
加拿大魁北克

摄影师：
Marc Cramer

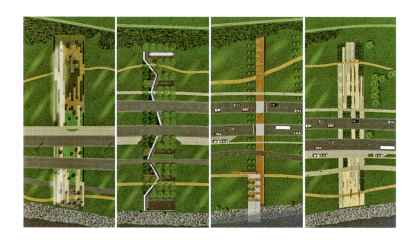

The park is a 2.5km long river bank, located in Quebec. It is built in honor of 400 anniversaries of Quebec City, designed by the designer Daoust Lestage who is designated by the government.

公园位于魁北克是一个长 2.5 公里的滨水河岸，是为纪念魁北克市成立 400 周年而建立的，由魁北克市政府任命设计师 DAOUST LESTAGE 来设计。

With single lane, walkway and parking lot, leading to St. Lawrence River, the New Park creates a public riverside area for the citizens in Quebec City. The total budget of this Park reaches up to USD 75 million, while the actual cost is only USD 86 million when completed. The walkway on the dock is chosen as the starting point of the Park and becomes the important entrance of this Park. The riverside dock and its structure are reconstructed; a landmark - the new beacon is designed and becomes the new view of this desolate industrial area.

这个有单车道、散步道和停车场，并能通往圣罗伦河的新公园为魁市的市民开创了一个滨水公共活动场所。该公园总预算达 7500 万美元，而竣工时实际只花了 8600 万美元。工程起点选址在一个码头散步道上，并成为进入公园的重要入口。建筑师对这个河边码头及其结构重新进行了改造，设计出一个地标性的灯塔，使其成为这个被工业废墟覆盖的场地的新景观。

The architectural design is inspired by the land features and the logging and shipbuilding industry. The reconstruction of the site and the design for the coastal landscape are deeply rooted in the memory of the citizens in Quebec and the stories of this important historical period, and also form the features of the whole project.

这种建筑设计是受场地地形特点与 19 世纪木材砍伐与造船业的启发而来的，对基地的改造和沿海景观设计都深深植根于魁北克市民的记忆与这段重要历史时期的故事，而这也构成了整个工程的特征。

布法罗河口市政公园
Buffalo Bayou Municipal Park

LOCATION:
Houston TX
CLIENT:
Buffalo Bayou Partnership
TEAM:
Tim Peterson
Scott McCready
Lance Lowrey
Rhett Rentrop
John Brandt
Kevin Shanley, FASLA
Nancy Fleming

项目地点：休斯敦 得克萨斯州

客户：
布法罗巴尤伙伴关系

团队：
彼得森·蒂姆
斯科特·麦克里迪
兰斯·洛雷
Rhett Rentrop
约翰·勃兰特
凯文·尚利，FASLA
南希·弗莱明

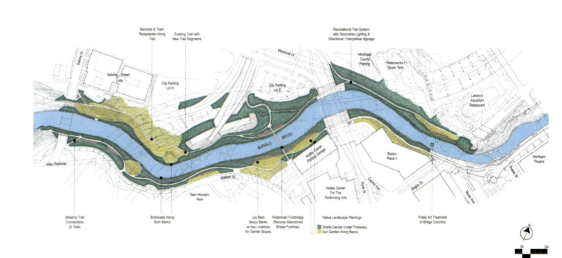

黄　蓝

对林　暖色铺装　白色硕石　低矮灌木　水边树池　水中

Since arthur comey did his city plan for Houston in 1912, people have talked about making the city's bayous into linear parks, but it was not until the 1970s and 80s that pieces began to fall into place. The landscape architect had been hired by the Buffalo Bayou partnership to provide an early conceptual master plan transitioning urban bayou treatments to pastoral bayou treatments east and west of the downtown, which had been encircled by freeways and arterials.

自从亚瑟戈美在 1912 开始为休斯敦市做规划，人们都在谈论使城市的支流成为线性公园，但直到 20 世纪 70 年代和 80 年代才开始着手。景观设计师曾参与过布法罗河口的早期规划，由他来将概念性总体规划城市转为现实，以田园河口规划改造已被包围了的东西部高速公路和干道。

福冈堰樱花公园
Fukuoka-Zeki Sakura Park

LOCATION:
Ibaraki Prefecture Japan
AREA:
2.7 ha
DESIGN COMPANY:
Keikan Sekkei Tokyo Co.,Ltd.

项目地点:
日本 茨城县
面积:
2.7 公顷
设计公司:
Keikan Sekkei Tokyo Co., Ltd.

Fukuoka-Zeki Sakura Park is located along the Kokai River near Fukuoka-Zeki (Fukuoka dam), one of 3 major historic dams in the Kanto region. This project was a collaboration effort between Ibaraki prefecture and Tsukubamirai city, to commemorate the merger of the town of Ina, and the village of Yawara, which together form Tsukubamirai city.

福冈堰樱花公园坐落于福冈堰 (福冈坝) 附近的 **Kokai** 河畔。福冈堰是 **Kanto** 地区最具历史意义的三大堤之一。这是茨城县和筑波未来市的合作项目,以纪念组成筑波未来市的伊奈县与谷原县的合并。

The park maintains and fully utilizes the vernacular landscape around the historic Fukuoka dam while paying careful attention to the ecological function of the site. Existing woods were preserved to the fullest extent possible in order to maintain the biodiversity around the site.

公园保持且充分利用了具有历史意义的福冈大坝周围的民居景观,同时高度重视此地区的生态功能。为了保持该地区的生物多样性,现存的森林被进行了最大程度上的保护。

Sakura(Cherry) theme was celebrated in the park concept, because the site was originally famous for its beautiful cherry promenade. The "Cherry and Wind" monument welcomes people at the main entrance, symbolizing the celebration of the birth of Tsukubamirai City.

公园举办以 **Sakura** (樱花) 为主题的庆祝活动,这是因为该地区有史以来就以美丽的樱花漫步而闻名。中心入口的"樱花与微风"纪念碑,欢迎着人们的到来,同时象征着欢庆筑波未来市的诞生。

Providing an amenity where people can socialize and connect with nature and water was one of the important goals for this project. "Water Monument", "Mist Fountain" and "Jabu-Jabu Pond" are central park elements providing visitors with highly interactive water activity during summer.

这个项目最重要的目标之一就是要给人们进行社交及亲近大自然和水提供便利。而水之源纪念碑、雾泉和 Jabu-Jabu 人工水池都是公园的中心设施，它们给游客提供了可在夏季有高度互动性的水上活动。

Seasonal changes such as cherry blossoms in spring, water amusement in summer, and fall color in autumn add year round interest to the park.

四季的变换，如春天的樱花绽放、夏日的水上嬉戏，秋日的落叶景观让公园一整年都充满了乐趣。

德累斯顿动物园的非洲平原——长颈鹿与斑马园

African Plains In The Dresden Zoo — Giraffes And Zebras

LOCATION:
Bavarian German
AREA:
0.35 ha
DESIGN COMPANY:
Landschaftsarchitekten with Heinle, Wischer und Partner. Dresden

项目地点：
德国 巴伐利亚州

面积：
0.35 公顷

设计公司：
LANDSCHAFTSARCHITEKTEN WITH HEINLE, WISCHER UND PARTNER. DRESDEN

The imposing scenery of the park "Großer Garten" is used as a visual extension of the zoo and so gives the giraffe barn the impression of vastness. In elongation of the existing visitor ways there are headlands extending in the outdoor enclosure. That way the visitor gets the impression of being among the animals. The several zones of the barn are assigned with topics like savannah, water hole or scrubland to show the giraffes in their natural habitat. A scout-tree gives the opportunity to watch the animals at eye level.

长颈鹿园有大花园令人难忘的风景作为视觉延伸，给人空间广阔的印象。除了传统的游览方式外，我们在户外设计了很多瞭望台，通过这种方式让游客感觉仿佛置身于动物之中。园区用几个不同主题来展现长颈鹿的自然栖息地，例如草原、水域、灌木丛。瞭望台提供了从人的视线高度观赏动物的机会。

For those who promenade in the park "Großer Garten" the old zoo entrance is revived to the new "Zoo window".There are small keyholes in the portal where you can look through and get appetite of a visit in the "Zoo Dresden".

对于那些在景观大花园中散步的人来说，动物园的老入口现在已经成为展现动物园风采的新窗口。公园入口大门上有很多小孔，通过这些观察孔人们可以了解到动物园里面的情况，使他们产生参观德累斯顿动物园的欲望。

西首尔湖公园
West Seoul Lake Park

LOCATION:
Seoul Korea
LEAD DESIGNER:
Sehee Park;Chihun Kim, Hyejin Cho, Youngsun
Jung, Saeyoung Whang, Miran Lyu
LANDSCAPE DESIGN:
Prof. Byunglim Lyu
TEAM:
Junsuk Bae, Changwon Lee, Seong ki Kim,
Taeyoung Ko, Sangkook Lee, Dongwon Kim,
Suhyun Kim, Jungun Choi, Sanghun Yoon, Kwangho
Hong, Hyunjung Lee, Myungbo Son, Heejin Park,
Jihwan Kim, Semin Oh, Eunji Kim, Yoonyoung Lee,
Yoon Jang, Wonki Jang
DESIGN COMPANY:
CTOPOS

项目地点：
韩国 首尔
首席设计师：
**Sehee Park;Chihun Kim, Hyejin Cho, Youngsun
Jung, Saeyoung Whang, Miran Lyu**
景观设计师：
Prof. Byunglim Lyu
团队：
**Junsuk Bae, Changwon Lee, Seong ki Kim,
Taeyoung Ko, Sangkook Lee, Dongwon Kim,
Suhyun Kim, Jungun Choi, Sanghun Yoon,
Kwangho Hong, Hyunjung Lee, Myungbo Son,
Heejin Park, Jihwan Kim, Semin Oh, Eunji Kim,
Yoonyoung Lee, Yoon Jang, Wonki Jang**
设计公司：
CTOPOS

Our core design concepts were regeneration, ecology, and communication.
We intended to redesign the boundary, which had disconnected park and local
community. Approach to the park would be easier and our intervention would foster
connectivity with local residents.

我们的核心设计理念为再生、环保及沟通。阻隔了公园和社区的旧边界将被重
新设计。新公园将更方便进入，会促进当地居民的交往。

First, the park was created as an "open cultural art space," embodying the diversity of the area's identities and urban cultures, using the native environment to foster a self-organizing cultural zone for everyone.

首先，公园已经被打造成"开放的文艺空间"，既体现出地区特色和城市文化的多样性，又利用当地环境为公众打造以自我组织为特点的文化圈。

Second, the park preserves the existing natural topography and scenery to create a space for "urban ecology", integrating nature, culture, and urbanity. We engage environmental elements of the site to make a cultural space for events with a home-grown culture—a scene open to all.

其次，为了结合自然、文化和都市风格，打造"城市生态"空间，园区保留了现有的自然地形和景观。我们利用该地区的环境因素，为本土文化活动打造空间，并向公众开放。

Third, it was created as a "people's park," uniting visitors by featuring a variety of abundant park events and special programs. This park is a citizen's park, citizen-generated by participation and communication. The school of ecology education will teach people the value of water and forest of the natural environment, scenery conservation, and nature study. Our use of program and context encourages all classes to communicate with one another.

第三，公园将以各种丰富的活动和特色节目聚拢游客，使之成为名副其实的"人民公园"。它是属于老百姓的，在其间每个人都可以参与和交流。该公园还是开展生态教育的学校，它告诉人们自然环境中的水和森林，风景保护和自然研究的价值。我们的项目和环境鼓励不同社会阶层的沟通。

Fourth, as a former water treatment plant reborn as a "site for urban culture," materials from the old plant were reused in surprising, inventive ways to transform raw nature into a new eco-functional space.

第四，公园的前身是一家净水厂，现在脱胎换骨成为"城市文化基地"，旧厂房的材料竟然被重复利用，创造性的方法将原始的自然改造成为新的生态功能空间。

晋城市儿童公园
Kinderpark Jincheng

LOCATION:
Jincheng Shanxi
AREA:
5,1 ha
PHOTOGRAPHY:
Rehwaldt Landschaftsarchitekten
AWARDS:
Excellent Landscaping Project Gold Prize,China Society of Garden Landscape in 2010
DESIGN COMPANY:
Rehwaldt Landschaftsarchitekten,Dresden,Deutschland

项目地点:
山西 晋城

面积:
5,1 公顷

摄影:
雷瓦德景观设计事务所

获奖:
优秀园林绿化工程金奖，中国园林风景协会 2010 年

设计公司:
德国德累斯顿雷瓦德景观建筑设计事务所

The functional and formative trisection of the area in north south direction was one of the main design ideas.

基本思路是根据设计和功能上的特点将园区由南至北分为三个区域。

The south has an urban character and reflects modernity. It became a distinctive place as contemporary entrance and representation space.

南面的区域具有城市特点，并且反映了现代特点。这一区域，作为具有现代特点的入口空间以及得到着重表现的空间，为公园创造出独特的"地址"。

The north is more landscaped and emblematizes tradition. This area is a recreation zone.

与之相对应，北面区域更多体现了自然风格，象征着传统。这里的主要功能是供人们休息。

The interspace – the exhibition and event zone in the park center – is intended to convey between urbanity and landscape. Therefore, design elements of both parts were set. And in content, information boards on urban development and modern works of art were aligned.

中间的区域——展览及活动区——介于城市风格与自然风格、现代与传统之间。此区域的主题由展览的内容决定（介绍城市发展的信息墙和现代艺术品），形式上通过两个区域中的设计元素来实现。

Structures like the Bonsai garden and the Muslim cemetary which were worth retaining were integrated into the new design concept.

需要留的部分，如盆景园和回民墓，将融合在设计之中。

Besides the main entrance in the south there are now more park entrances in the east and north.

除了南面主入口，东面和北面也设有公园入口。

The Jincheng surroundings gave the inspiration for the new park design – islands, boardwalks, colourful and species-rich forests, lakes. In form, any natural landscape in China gave the model for the landscape park. The human influenced agricultural landscape with its organic shapes in Jincheng region was used to design the exhibition area. The idea for the urban plaza derived from the aligned Jincheng urban structures. Three different park zones emerged by transfer the motives onto the park site plan.

公园设计的出发点取自于晋城的周边环境——岛屿、栈桥、颜色与植物种类丰富的森林、湖泊。所有的中国自然景观都可以成为这座景观公园在形式上的范本。晋城当地人为形成的起伏的农业景观激发了自然风格与城市风格之间展览区的设计思路。城市广场的设计思路则是来自于晋城的城市设计。通过公园中不同区域之间的过渡形成了三个不同的公园区。

Jon Storm 公园
Jon Storm Park

LOCATION:
Oregon USA
DESIGNER:
Lango Hansen Landscape Architects
DESIGN COMPANY:
Lango Hansen Landscape Architects Firm

项目地点：
美国 俄勒冈

设计：
Lango Hansen Landscape Architects

设计公司：
Lango Hansen Landscape Architects Firm

The design of Jon Storm Park provides opportunities for dramatic views of the Willamette River, lawn areas for flexible use, and trails that wind through riparian habitat. The park provides a number of walkways that provide direct connections to the parking lot, restroom, transient walkway and a cantilevered overlook. The cantilevered overlook dramatically overhangs the existing sheet pile wall and offers views of the Willamette Falls and the river below. Adjacent to the main walk, interpretive signage relays the history of the area and the importance of this open space to the development of Oregon City.

在 Jon Storm 公园可以欣赏到 Willamette 河的壮丽景观。公园里设有草坪区，可以用于各种用途；小径在河边蜿蜒伸展。公园里有许多人行道，可以直接通往停车场、休息室、过渡人行道和一个悬空的看台。悬空看台从板桩墙伸出，在上面可以欣赏 Willamette 瀑布及下面河流的壮观景色。主通道附近的标识牌传递着这个地区的历史气息，同时也反映出这片空地对俄勒冈发展的重要性。

The parking for the park is located on ODOT property underneath the I-205 overpass. Parking for 30 spaces are provided along with bioswales planted with native grasses that filter drainage off the asphalt surface before it enters the river. ADA spaces are provided in the parking area and all walks within the park meet accessibility requirements.

公园的停车区安排在俄勒冈州运输部（ODOT）的地产上，位于 I-205 立交桥的下面。30 个停车位沿着种满本地草类的生态沼泽地布置，生态沼泽地可以在沥青路面排出的污水进入河流之前将其过滤。停车场内设置了符合美国残疾人法案（ADA）的停车位，公园内所有的通道都符合方便进出的要求。

In addition to picnic tables within the grass areas and adjacent to the walkway, a number of stone walls have been constructed which provide seating areas and also recall the historic stone walls that are found throughout Oregon City. These walls provide areas for people to gather and enjoy the view, eat lunch and watch the riverboats along the river. A picnic shelter is also located near the overlook.

除了草坪区和人行道附近的野餐餐桌以外，公园内还设置了许多石墙，可以用作座位区，使人联想到了俄勒冈市随处可见的历史石墙。有了这些石墙，人们就可以聚在一起欣赏风景，吃午餐，或欣赏河边的河船。看台附近还有一个野餐屋。

As part of the park improvements, the City repaved Clackamette Drive which is the major street adjacent to the park and constructed a large turnaround across from the park. This was done for the City's trolley and tour buses which make their way down this street that has no outlet.

作为公园改造的一部分，俄勒冈市重新铺设了公园附近的主干道 Clackamette 车道，并且在公园对面修建了一条宽敞的回车道。这条街道本来没有出口，修建回车道以后，城市里的电车和旅行车就可以从这里通过。

The improvements at Jon Storm also included a trail called the Willamette trail which connects Jon Storm Park to the existing Clackamette Park. This multi-model asphalt trail is 12-feet wide and winds through areas that were replanted with native vegetation to satisfy permitting requirements.

Jon Storm 公园的改造还包括一条名为"Willamette 小径"的通道，这条小径把 Jon Storm 公园和原来的 Clackamette 公园连接在一起。这条多模式的柏油小路宽 12 英尺，在种满本地植物的地面上蜿蜒伸展，以满足各种要求。

南森公园 奥斯陆
The Nansen Park Oslo

LOCATION:
Oslo Norway
AREA:
200 000 m²
LANDSCAPE ARCHITECT:
Bjørbekk,Lindheim

项目地点:
奥斯陆 挪威

面积:
200 000 平方米

景观设计:
Bjørbekk & Lindheim

An old cultivated landscape with much variation and beauty was levelled into Oslo's international airport in the 1940 - 60s. In 1998, the airport was moved and left behind it a peninsula of almost 404.7ha in need of transformation.

20 世纪 40 年代到 60 年代，为了建造奥斯陆国际机场，这里变化多样的秀丽培植景观被夷为平地。1998 年机场搬走以后，留下一个近 404.7 公顷的半岛，急需改造。

The moving of the Oslo International Airport at Fornebu resulted in the largest industrial reclamation project in the country. The new park was to form a functional focus and an identifying centerpiece of a new community some 10 kilometres from downtown Oslo. Plots for housing and offices were sold off to private developers, while the Norwegian Directorate of Public Construction and Property together with the City of Oslo undertook responsibility for infrastructure and landscape; the treatment of polluted grounds, and the planning of a new park structure.

Fornebu 奥斯陆国际机场的搬迁使这里的改造成为国家最大的工业垦殖项目。新建公园主要以功能为主，形成一个地标，距离奥斯陆市区约 10 公里。用于住宅和办公的地块卖给了个人开发商，而基础设施和景观的建设、污染地面的处理以及新公园结构的规划则由挪威公共建筑和财产管理局和奥斯陆市负责。

In 2004, an architectural competition was won by landscape architects Bjørbekk & Lindheim.

2004 年，景观建筑师事务所 Bjørbekk & Lindheim 在本项目的建筑竞标中中标。

The central Nansen Park (approx. 200,000 m²), has been designed to serve as an attractive and active meeting place for all those who will be living at Fornebu. A strong identity, simplicity and timelessness have been key points. In order to respond to its dramatic history, the park has been designed as a dynamic dialogue between the uncompromising linearity of the airport and the softer, more organic forms of the original landscape. The site borders the Oslo Fjord on three sides. The openness of the landscape, as well as the distant contours of the hills gives a strong and peaceful feel of the sky, a separateness and spaciousness which we have tried to instil in the new landscape. The quiet calm of the extensive views and the harmonious forms have carefully been combined with activities.

南森公园中心（约 200 000 m²）将设计成一个充满活力并且具有吸引力的场所，供 Fornebu 市民聚会之用，主要强调特性、简约和永恒。为了与其生动的历史相呼应，公园在机场生硬的线性和原有景观柔和、有机的构造之间形成生动的对话。这里三面都与奥斯陆峡湾相邻，景观的开敞性和远处小山的轮廓给天空带来一种强烈的平和感，一种我们一直努力向景观中灌输的超脱感和宽敞感。广阔的视野形成的宁静与和谐的形态与功能区巧妙地融合在一起。

A strong ecological profile forms the foundation of the whole transformational process. Polluted grounds have been cleaned, asphalt and concrete have been retrieved and reused, new soil for cultivation was made from masses from the site. Large volumes of earth and rock within the Fornebu area have been used to transform the flat airport area into a landscape with different spatial qualities and heights to create views towards the fjord. Engineering firm Norconsult and the German firm, Atelier Dreiseitl was consulted during the planning phase.

强烈的生态概貌成为整个改造过程的基础。受污染的土地已进行清理，沥青和混凝土都进行了回收和再利用，用于培植的新土壤从现场的土块中获取。Fornebu 地区的大块土和岩石把平整的机场用地打造成了具有不同空间特质的、高度不同的景观，面对着海峡。工程公司 Norconsult 和德国公司 Atelier Dreiseitl 都参与了项目的规划。

Pilestredet 公园
Pilestredet Park

LOCATION:
Oslo Norway
AREA:
70 000 m²
AWARD:
The City of Oslo Architecture Prize (Oslo Byes
Arkitekturpris) in 2005 and the National Building
Design Prize(Statens Byggeskikkpris) in 2007.
LANDSCAPE ARCHITECT:
Bjørbekk & Lindheim

项目地点:
奥斯陆 挪威

面积:
70 000 平方米

完成时间:
2005 奥斯陆城市建筑奖   2007 国家建筑设计奖

景观设计:
Bjørbekk & Lindheim

Pilestredet Park is an urban-ecology pilot project. When the old national hospital in Oslo moved to a new site more than 7 ha were converted to a residential and recreation area in the middle of town. Pilestredet Park is a car-free oasis in the city center with vehicles largely directed outside an area that is designed to accommodate the needs of cyclists and pedestrians. Surface water drainage and storm water management characterise the facility and exploit the natural 16 meter fall of the site. There are rippling streams, water canals and pools in all outdoor areas. Every drop of water is taken care of and used several times to trickle, flow and drip, or lie perfectly still and reflect the sky and the treetops.

Pilestredet 公园是一个城市生态学试验项目。奥斯陆原国立医院迁址之后,城镇中心多于 7 公顷的地区都变成了居住和娱乐区。Pilestredet 公园是市中心一片没有汽车的绿洲,车辆大部分都安置在为骑脚踏车和行人设计的区域之外。地表水系和雨水管理体系以其设施为特色,并充分利用了场地 16 米的自然落差。室外有潺潺的小溪、水道和水池。每滴水都经过精心的处理并反复使用,流淌、滴落或静静地躺着,反射着天空和树梢。

The project is based on the environmentally friendly principle of recycling building materials and elements from the old hospital. Stairs, foundation walls, window frames and granite gates have been preserved and reused in flooring, stairs and edging. The venerable portals have been used in the climbing wall or reset as part of the frame for the sandpit and pools. Concrete and other building rubble has been crushed and used for refill and as an aggregate in cast concrete used in the roads and public spaces. The terms "rag rugs"and "patchwork" from the world of textiles are thus directly transferred in the making of outdoor flooring in Pilestredet Park.

本项目遵守环境友好原则，使用了原医院的建筑材料和构件。原医院的楼梯、基墙、窗框和花岗岩大门都被保留了下来，重新利用到地面、楼梯和边缘的建造中。古老的入口门被用作攀岩墙，或作为沙坑或水池框架的一部分。混凝土和其他建筑石块都进行了破碎处理，然后重新用作回填或道路和公共空间使用的浇筑混凝土的骨料。在 Pilestredet 公园中，纺织业的术语"碎呢地毯"和"拼凑"被直接应用在室外地板的建造之中。

The old pool has been restored and a new element added, a steel frame with a water curtain feeding water in to the basin. This has become a popular playing area for children, as has the large arrow-head snake that children use in all sorts of ways.

旧水池进行了修复，并增加了一个新的元素，即一个带水幕的钢架，向池中注水。这里有一个大的箭头形弯道，孩子们可以以各种方式玩耍，因而成为深受孩子们喜爱的嬉戏场所。

When the entire hospital moved the site was left with majestic, leafy trees and extra effort has been invested in retaining these even in circumstances where they are very close to buildings. New trees have been planted to replace the old ones when they eventually die and a lot of bushes, creepers and ground-cover plants have been added to create a lush, green area in the center of town. Ground-cover plants also reduce the need for constant attention to weeding and facilitate maintenance.

医院搬走以后，留下了茂盛、雄伟的大树。即使这些大树距离建筑很近，我们也设法将其保留下来。如果老树死掉了，我们会重新种植新的树木，同时还种植了许多灌木、爬行植物和地被植物，在市中心打造出郁郁葱葱的绿色景象。此外，地被植物方便维护，无须经常除草。

The old National Hospital built in 1883 was surrounded by a massive wall to protect it from the public; this has been placed under heritage orders but punctuated in several places to give better access to the new area. Once inside there are no fences or restrictions and the public can move freely from one outdoor space to another. These public spaces are used by many as a pleasant green oasis on the way to and from work, whereas others use it as a place to escape from the turbulence of the city, a place to peacefully eat their lunchtime sandwiches or relax in a peaceful, green environment whilst still in the center of town.

原国立医院建于 1883 年，厚重的围墙将其与公众隔离。遵照遗产保护的指令，这些都被保留下来，不过有些地方也做了一些改动，以方便人们进入新建的区域。进入之后就没有围墙或限制了，人们可以从一个室外空间自由地进入另一个室外空间。很多人都将这些公共空间当作上下班途中舒适的绿洲，而另外一些人则将其作为远离城市喧嚣的场所，或是可以安静地吃午餐三明治的场所，抑或是在市中心的一处平和、绿色的放松场所

圣·安娜区公园
S. Anna Park

LOCATION:
Lucca Italy
GARDEN AREA:
22 000 m^2
PHOTOGRAPHER:
Gianfranco Franchi
DESIGNER/ARCHITECT:
Gianfranco Franchi

项目地点：
意大利 卢卡

面积：
22 000 平方米

摄影师：
Gianfranco Franchi

景观设计：
Gianfranco Franchi

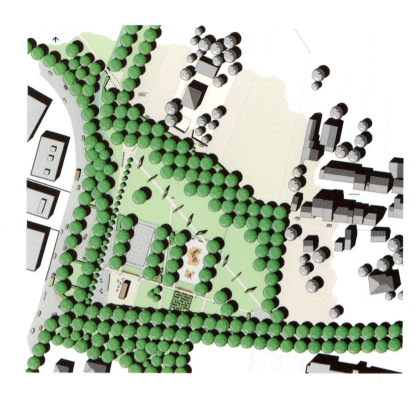

The park is found in the suburbs of the city of Lucca, in the S. Anna district. The park has the function of limiting the borders of the city and constituting a filter with the surrounding countryside. It is collocated between a Shopping Center and a residential neighbourhood.

公园位于圣·安娜区卢卡市的市郊，不仅界定了城市的界限，而且还形成附近村庄的过滤器，成为介于购物中心与住宅小区之间的一种公共空间。

The park is part of a network of parks linked to the Serchio river, which surrounds the city.

公园是与塞尔基奥河相连的公园网络的一部分，将城市围绕起来。

When we projected the park we tried to understand how we could create a socially useful space while keeping its historical memory alive.

在对公园进行设计时，我们试图找到一种方法：既能打造一个具有社交功能的空间，同时还能保留其历史意义。

We wanted to characterise the new park as if it had always been there in the mind of the visitors. Then we tried to understand which project and which design could be used and which materials and shapes would best suit the environment.

我们希望使新公园对于游客来说看起来像是与生俱来的一样，于是我们尝试着寻找可以借鉴的项目和可以使用的方案，以及最适合这个环境的材料和造型。

We decided to use a geometric mesh and poor materials such as concrete and corten steel, considering other shapes not appropriate for what we wanted to do.

鉴于其他造型不适合我们所希望打造的环境，我们决定采用几何网状的造型及混凝土和低合金高强度钢等材料。

The project redevelops a part of the countryside and the use that has been done of the trees recalls the typical cultivations of the surrounding environment.

本项目重新开发了乡村的一部分，而对于树木的利用则唤醒了对周围环境的典型塑造方式。

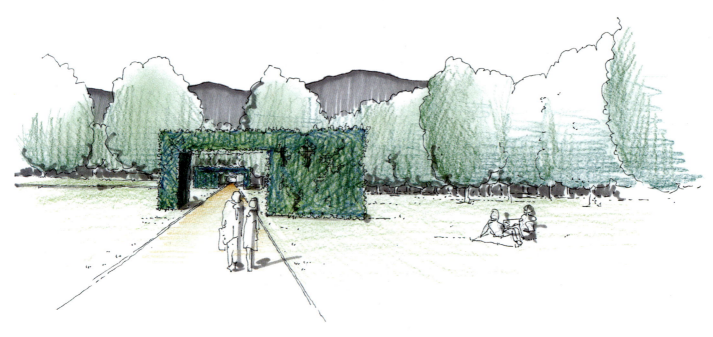

The park is characterised by small squares, by artificial hills of geometric shape, and by various playgrounds and spaces where relax and rest.

小广场、具有几何造型的假山和各种运动场以及放松和休息空间成为公园的特色。

A path which passes through the entire park, from East to West, is characterised by metallic structures where, during spring time, different types of plants flourish magnificently.

一条小路从东向西贯穿整个公园，小路旁边设置了一些金属构筑物，为小路增加了特色。春天的时候，各种植物在这里竞相生长。

Big metallic vases recall the traditional use of clay pots in the countryside.

大金属花瓶不禁使人们想起以前乡村地区经常使用到的　土烧制的花盆。

The structure of the vegetation is characterised by Populus Alba and Populus Nigra: typical cultivations of the environment.

公园主要采用白杨和黑杨进行绿化，这两种植物是打造空间环境的典型植物。

悉尼 Pirrama 公园
Sydney Pirrama Park

CLIENT:
City of Sydney Council
LOCATION:
Former Water Police Site, Pirrama Rd, Pyrmont
LANDSCAPE ARCHITECTS / LEAD CONSULTANTS:
ASPECT Studios Pty Ltd.
PROFESSIONAL PHOTOGRAPHERS:
Florian Groehn, Adrian Boddy

甲方：
悉尼市政府
地点：
悉尼 Pyrmont 区 Pirrama 路原水上警察中心
景观设计（主要顾问）：
澳派（澳大利亚）景观规划工作室
专业摄影师：
Florian Groehn，Adrian Boddy

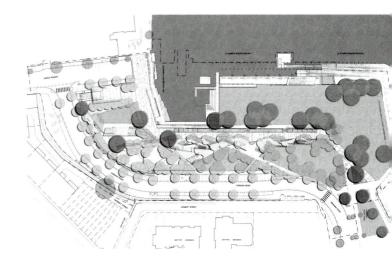

ASPECT Studios was commissioned by the City of Sydney to design a new waterfront park on the former water police site in Pyrmont. The brief was to develop a master plan for a 1.8 hectare parcel of land on the Pyrmont peninsula into public parkland incorporating a significant children's play environment.

在悉尼市委员会的委托下，澳派(澳大利亚)景观规划工作室在地区的水上警察中心的旧址设计一个新的滨水公园。设计任务书要求把半岛上 1.8 公顷的土地发展成为一个都市公共公园，并把公园打造成为一个可供儿童游乐的城市空间。

The New Park on the Former Water Police site required significant marine engineering at the harbour edge to create a sheltered bay and interpret the former shoreline.

悉尼海滨公园的设计对海洋工程进行了详细的考虑，把海湾打造成为一个舒适而又高使用率的新城市空间，并且把水上警察的历史风貌保留下来融合到新的景观之中。

The public realm includes wharfs, promenades, squares, laneways, rain gardens and a cycle way which forms significant public fabric, linking the City to the Docklands.

悉尼海滨公园设有海港码头、滨海大道、城市广场、小巷回廊、雨水花园、自行车道等，形成一个充满活力的城市肌理空间，增强了城市与滨海空间的联系。

The bay creates passive recreation opportunities at the water's edge and strengthens the site's historic relationship to Sydney Harbour. A range of other "park rooms" are created which celebrate this unique location.

海滨公园的打造为悉尼市民提供了一个滨海休闲活动与聚会的空间，并突出体现出基地与悉尼海港的历史联系。由于海滨公园位于悉尼市中心的黄金地段，不同类型的花园空间的打造，保证了海滨公园的功能性与特色。

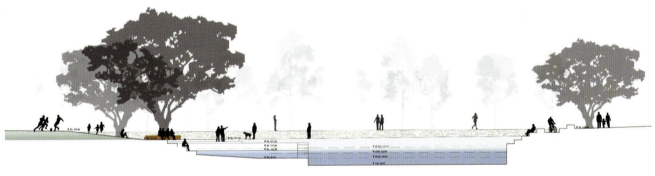

World's best practice initiatives were embedded into the master plan and rain gardens and bio-filtration trenches in the park capture and clean the water from the surrounding park storm water catchment. Street tree pits along Pirrama Rd collect street runoff and 200,000KL water tanks have been proposed to ensure irritation is maintained sustainably throughout the year. Add to that, the proposal of solar panels on the shade canopies to power park lighting and the master plan is an exemplar of best practice ESD. Social sustainability is promoted through the creation of a significant public space at the end of Harris Street which provides an opportunity for social interaction, public gathering and displaying community wealth.

在公园的总体规划设计中融入一系列世界上最领先、最优秀的生态做法，即在景观设计中融入雨水花园和生物过滤槽，可以在公园内收集和清洁公园集水区的水体。大道的行道树种植槽用来收集街道的雨水径流，再用水箱储备200,000千升的水，保证全年的灌溉用水。此外，在公园的遮阳篷上设太阳能电池板，收集太阳能，为公园提供照明，总体规划体现了环境可持续发展的最佳范例。在街尽头建立一个重要的公共空间，有利于人们的互动交流和公众集会，展现社区生活美好的一面。

Lørenskog 的 Kjenn 市政公园
Town Hall Park at Kjenn in Lørenskog

LOCATION:
Akershus Norway
AREA:
39 000m²
LANDSCAPE ARCHITECTS:
Bjørbekk & Lindheim

项目地点：
挪威 阿克斯胡斯
面积：
39 000 平方米
景观设计：
Bjørbekk & Lindheim

The park is in constant use with nursery schools and school children using it during winter days for skating and tobogganing on the ice rink and in summer for concerts, picnics, fly fishing courses, fishing, kiting, remote controlled boats and airplanes, feeding of ducks and so on.

托儿所经常使用这个公园。冬天的时候，孩子们可以在溜冰场滑冰或者乘雪橇滑雪；夏天的时候，孩子们可以在这里举行音乐会、野餐、参加飞钓课程、钓鱼、放风筝、玩遥控船和遥控飞机，或喂鸭子等活动。

The park is located close to the town center in beautiful natural surroundings close to the Town Hall, to Lake Langevann, to the new Mainland High School and also to Kjenn Junior School. Lørenskog municipality lies to the north of Oslo and is about a15min commute. It is poised to carry out a comprehensive downtown expansion in which the large, new arts center, "Lørenskog House", will be an important element. The new park is linked to the new Lørenskog center via a new, soon to be opened pedestrian bridge over Route 159.

公园位于市中心附近，距离市政厅、LANGEVANN 湖、新大陆高中和 KJENN 初中都很近，周围自然风光秀美。LØRENSKOG 市政府在奥斯陆以北，来回大约 15 分钟。计划将进行一次全面的市区扩建，在此次扩建中， "LØRENSKOG 住宅" 是一个重要元素。新公园通过 159 大道上方即将开通的新建过街天桥与新的 LØRENSKOG 中心连接在一起。

With the new downtown development, the expansion of Ahus (Akershus University Hospital) and the jobs provided there and the establishment of new large mail terminal for the Norwegian postal services within a few miles radius of the park, there is also a need for attractive green spaces, footpaths and bike paths that bind it all together. The park at Town Hall will acquire a whole new status.

随着新市区的开发，Ahus（阿克什胡斯大学医院）的扩建及其提供的就业机会，以及距离公园几英里的挪威邮局大型邮政点的建设，这里还需要建设一些绿地、人行道和自行车道，使这些设施融为一个整体。市政公园将会具备一个全新的地位。

The park at the Town Hall is located in a well-established recreational area that has commonly been used for concerts, performances and as a recreational area for the city's inhabitants. Particularly in winter the sloping grassy site has been an attractive toboggan run.

市政公园位于一个设施齐全的娱乐区，这里通常会举行一些音乐会和表演，市民也可以在这里娱乐，尤其在冬季，在倾斜的草地上乘雪橇滑雪是一个不错的选择。

新 Waldkirchen 城市公园
New Waldkirchen Urban Park

LOCATION:
Bavarian Germany
AREA:
7.5 ha
LANDSCAPE DESIGNER:
Rehwaldt
DESIGN COMPANY:
Rehwaldt LA, Dresden

项目地点:
德国 巴伐利亚
面积:
7.5公顷
景观设计:
Rehwaldt
设计公司:
Rehwaldt LA，Dresden

This small garden festival is based on a decentralised exhibition concept. The main area of the garden exhibition is designed as a loop path from the core of the garden exhibition, the new Waldkirchen urban park, via old town to the characteristic areas in and near Waldkirchen.

本次小型花园节以分散展览理念为基础。花园的主展区设计成环形路线，从花园展的核心 —— 新 Waldkirchen 城市公园，途径老城区，到达 Waldkirchen 市内和附近的特色区。

As the main feature of the garden exhibition a new urban park was developed on the south-eastern edge of the old town, along brook Waeschlbach. Through reorganisation of existing streets, parking sites and a bus stop, a spacious entrance square to the urban park and the garden exhibition could be established. Here, all main events and activities take place. "Landschaftsbalkone" (scenic balconies) are offering special views into the surrounding landscape and are explaining them.

新城市公园位于老城区的东南侧，Waeschlbach 小溪沿岸，成为花园展的主要特色。对原有街道、停车区和公共汽车站进行重新规划之后，有必要在城市公园和花园站附近建一个宽敞的入口，所有的重大事件和活动都可以在这里举行。观景台提供了观赏周围景色的特殊视野，同时也对周围的景色进行了很好的诠释。

The Waeschlbach valley features a variety of scenic and historic sites like the natural landmark "Gsteinet", floodplain forests and hillside meadows along Waeschlbach, as well as some protected habitats in an urban context. Along the elongated meadows the manifold topographies and the diverse landforms were used to develop differentiated gardens and to provide great views into the landscape. Within the garden exhibition a new open space system was established which links the urban park and the adjacent residentials through new paths.Within the structural facilities and design focuses the Waldkirchen's characteristic location as part of the Bavarian Forest was respected. Therefore, the material wood plays an important role as functional and artistic element. The elements of the urban park as entrance square, urban promenade, water stair case, cherry gardens, pond and Waeschlbach as well as the spacious meadows with the "Gsteinet" are understood and designed as self-contained areas. Manifold views beyond the park borders enlarge the spatial impression of the park.

Waeschlbach 溪谷以各种景点和古迹著称，如大自然的地标"Gsteinet"，Waeschlbach 沿岸的泛洪区森林和山腰草地，以及城市中一些受保护的生境。设计利用长长的草地旁边多样化的地形打造出不同的花园，提供了绝佳的观景视野。花园展内建立了　个新的空地系统，通过新建的小路将城市公园与附近的住宅联系在一起。结构设施和设计焦点都尊重了 Waldkirchen 作为巴戈利亚森林一部分的独特位置。因此，作为功能和艺术元素，木材在这里发挥着重要作用。城市公园的元素，如入口广场、城市散步道、水景楼梯、樱桃园、水塘和 Waeschlbach，以及带有 Gsteinet 的宽敞草原都被融入设计之中，组成一个设施齐全的区域。公园另一侧，远处多样化的景观增加了公园的宽敞感。

An additional focus of the garden exhibition is the garden "Bellevue" along the loop path. It represents the city's character besides the other municipal open spaces as market-place, cemetery, urban park and sports fields. At the "Bellevue" the unique view at Waldkirchen was used and staged through a self-contained design. Waterbound paths frame the shrub and bush plantings and show the inside and outside areas. On a spatial view point a water fountain was placed which is playground and recreational zone in one. On historic site of Waldkirchen's water reserve the site was new interpreted by using instruments of contemporary landscape architecture.

花园展的另一个亮点就是环形路旁边的"Bellevue"花园，它代表着除了市场、公墓、城市公园和体育设施等城市空间以外的另一个城市特征。通过综合的设计，Waldkirchen 独具特色的景观在"Bellevue"重新上演。亲水小路与灌木相互呼应，显露出内部和外部区域。鉴于空间的因素，喷泉所在的位置既是一个运动场，又是一个娱乐区。作为一个以 Waldkirchen 的水资源为特色的古迹，场地通过当代景观建筑得到了重新诠释。

悉尼 5 号湿地市政公园
Sydney Wetland 5 Municipal Park

LOCATION:
Sydney Australia
PROJECT SIZE:
20 000 m²
PHOTOGRAPHY:
Simon Wood, Sacha Coles
LANDSCAPE ARCHITECT:
ASPECT Studios

项目地点：
澳大利亚 悉尼

项目规模：
20 000 平方米

景观设计：
Simon Wood，Sacha Coles

设计公司：
澳派（澳大利亚）景观规划设计公司

 ASPECT Studios were commissioned to design and document Wetland 5, the culmination of the chain of wetlands in Sydney Park. This existing area is the oldest parcel of land in Sydney Park and the only piece which remains on deep soil (rather than on fill). The condition of the wetland was undermining the environmental benefits of the wetland system. ASPECT Studios brought extensive site knowledge to the project, having completed the detailed master plan for Sydney Park in May 2006.

澳派受到悉尼市议会的委托，为悉尼公园湿地链 5 号湿地进行景观设计和施工图设计。项目现场是悉尼公园现存的唯一拥有深层土（而非回填土的）的地块，也是悉尼公园最古老的地块。现场年久失修，木材腐烂，水土流失。现有的湿地不能全面发挥其对环境的促进作用。通过对现场的详细了解和考察，澳派在 2006 年 5 月圆满完成悉尼公园的详细总体规划设计。

The scope for the project included design development and construction of the wetland and surrounds including pathways, retaining walls, seating and shade trees. The materials selected reflect the vision of the detail master plan which was to recognise and mark this place as the culmination and holding point for the greater parks wetland system. The design of the high quality, in-situ concrete walls work as a semi circular bracket or "frame" to the park water course. ASPECT Studios saw the opportunity to mark the upgrade of the park by inserting a simple and robust gesture which is of its time.

项目的设计范围包括湿地的深化设计和施工图设计，以及周围道路、挡土墙、座椅和遮阳乔木的设计。所有的材料都根据总体规划设计的风格来选择。设计需要综合考虑整个湿地处理链，同时要考虑将 5 号湿地作为湿地系统的蓄水池。现浇的高质量混凝土墙形成公园半围合蓄水通道。通过简洁生动的设计，让公园得到功能的提升。

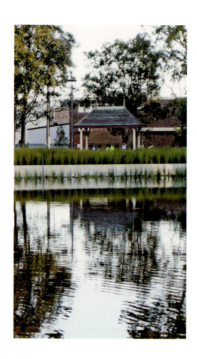

The concrete arc sets both the infrastructure and the organisational logic of the park and provides an informal seat and edge to the wetland. Fluorescent lights (which are triggered by a light sensor which respond to the environmental conditions) are housed in the wall creating a safe and usable space whilst also being an attractor at night.

湿地旁的弧形混凝土坐墙是公园的基础设施与水体流动通道。在混凝土坐墙内设有光线感应的荧光灯，保证夜晚公园的照明与安全。

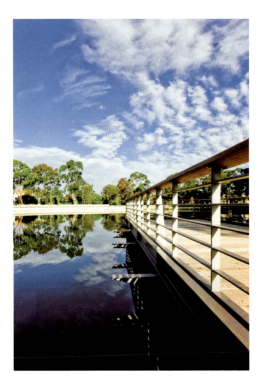

Cairnlea 公园
Cairnlea Park

CLIENT:
VicUrban
LOCATION:
Cairnlea Victoria
PROJECT DIRECTOR:
Rod Grist

客户：
VicUrban

地点：
Cairnlea Victoria

项目总监：
Rod Grist

"VicUrban" has been extremely pleased with the outcomes of the WSUD, water reuse and irrigation scheme, associated revegetation work and preservation of native grasslands within the Cairnlea Estate.

"VICURBAN" 非常满意整个设计成果：水敏城市设计、水资源再利用和灌溉方案、关联的植被重建以及对 CAIRNLEA 地产区内原有草地的保护工作。

The innovative approach taken by HASSELL in the conceptualisation, design and implementation of the landscaping works throughout the estate over the last 10 years has created a benchmark for future VicUrban projects in the development of sustainable master planned communities.

在过去的十年间，HASSELL 将创新体现在地产整个景观工程的概念、设计和施工方面。这为日后以叮持续发展为蓝图的城区建设中其他的 VICURBAN 项目树立了基准。

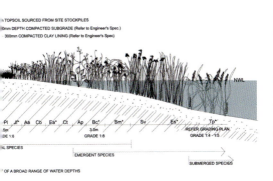

We put our full support behind HASSELL, as lead consultant for these works, in submitting this project for the Land Management Award in this year's Australian Institute Landscape Architecture Awards.

HASSELL 是这些项目的首席顾问，我们对其全力支持，并呈交此项目作品参与澳大利亚建筑委员会奖的土地管理类奖项的评奖。

埃布罗河 U12 市政滨河公园
Ebro U12 Municipal Riverside Park

LOCATION:
Zaragoza Spain
AREA:
87,535 m²
DESIGNER/DESIGN COMPANY:
Antonio Lorén, Eduardo Aragüés, Raimundo Bambó
PHOTOGRAPHER:
Aitor Ortz
PHOTOGRAPHER:
ACXT Architects

项目地点:
西班牙 萨拉戈萨

面积:
87535 平方米

设计师 / 设计公司:
Antonio Lorén, Eduardo Aragüés, Raimundo Bambó
(ACXT)

摄影师:
Aitor Ortz

设计公司:
ACXT Architects

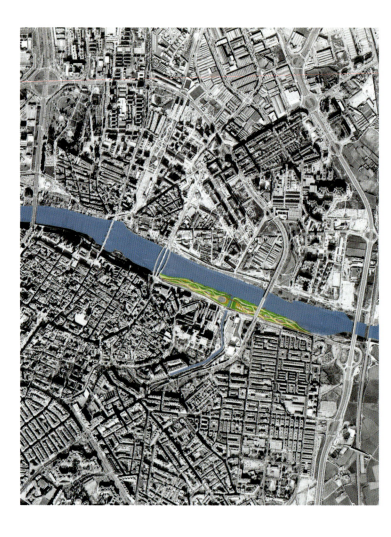

The north boundary of the Tenerías - Las Fuentes project area covered the space along the right bank between the river bed and the buildings that comprise the north edge of the Tenerías and Fuentes districts, running as far as the location of the future dam (calle Fray Luis Urbano) to the east and the Puente de Hierro (Iron bridge) to the west.

西班牙埃布罗河（Tenerías 至丰特斯段）滨河景观设计项目北临埃布罗河右岸，包括一条由南向北汇入埃布罗河的 Huerva 河，向南一直延伸到 Tenerías 至丰特斯区北部的建筑物。东至规划中的弗赖路易斯乌尔巴诺水坝，西至铁桥。

This is a built-up area consisting of the Echegaray and Caballero promenade and a park running lengthwise between this avenue and the river Ebro.
The possibility of creating a walkway along the right bank of the Ebro through a garden area running from the iron bridge along Echegaray promenade to near the bridge on the third ring road, taking into account the future Ebro dam pedestrian link, makes this project an outstanding route for pedestrians through the city.

此项目涉及很多建筑物，包括埃切加赖和骑士两条步行道，一座从此处一直延伸到埃布罗河的公园。沿埃布罗河右岸将建造一条人行道，从埃切加赖步道周边的铁桥出发，一直到三环的一座桥。这条人行道也考虑到了通往联接埃布罗大坝的线路，因而是一条绝佳的穿越城市的步行路线。

A municipal sports center near the mouth of the river Huerva, nestling between the river itself and Union bridge, benefits and encourages the park's future activities, more closely associated with open-air sports.

邻近 Huerva 河口，在埃布罗河和联合大桥之间，建有一个市立运动中心，有利于人们日后在公园里进行各类活动，尤其是户外运动。

The high-density population in the Fuentes district means there will be sufficient traffic and that the new routes will be used, which will encourage and promote the park's proposed activities.

丰特斯区人口密度人，需要充足的交通线路。因此新修建的道路会发挥作用，并有助于提高公园原计划中的各项活动。

The project creates a pedestrian link between the center of the city and the natural setting of Soto de Cantalobos and the river Gállego water park, and creates a network of routes running the length of the project area, clearly differentiating between the roads and paths for cars, bicycles and pedestrians in the city, and those for water, sports, leisure and pedestrian activities in natural settings.

此项目建立了一条沟通市中心、**Soto de Cantalobos** 自然景观和 **Gállego** 河滨公园的步行大道，打造了一个贯穿整个区域的公路网，不但将城市中汽车道、自行车道和人行道明确地区分开来，也清晰地展现出水景、运动、休闲和步行活动的不同位置。

The project enhances the view from this area towards the Basilica of Our Lady of the Pillar, creating a link between the project area and the city center thanks to one of its most famous landmarks.

通过这一项目，从这一地区向圣母大教堂眺望视野更开阔清晰。这样，作为市区地标建筑的大教堂就成为了项目区域和市中心之间的联系。

Earth excavated from the plot was relocated to modify the plot's topography in order to facilitate access to the water level. This remedied access problems whilst moving earth to create areas linked to the city level. Modifying the slope of the existing bank will, together with cleaning the riverside and clearing its shrubbery, allow unobstructed views and pedestrian access between the city and the natural setting.

施工人员重新开掘了项目区，使其地势更接近河流水平面。此举在解决了水岸问题的同时，也将开掘的土石填补到临近市区一侧，使滨河景观场地与市区水平高度一致。改造河岸坡地、清理河岸环境及岸边灌木植被，创造自由开阔的视野，也开辟了一条城市和自然景观之间的人行通道。

The mouth of the river Huerva would be enhanced by a new pedestrian walkway across its 60m span, whilst maximising communications between the riverbank roads of the Ebro and the Huerva, reinstating the river Huerva as a city thoroughfare and gateway to the Ebro.

在 Huerva 河的河口处将修建一条跨度为 60 米的人行步道，最大限度地连通埃布罗河及 Huerva 河之间的滨河道路，恢复 Huerva 河汇入埃布罗河主要通道的作用。

The Union bridge is envisaged as the main gateway to the Tenerías – Las Fuentes park. The new routes running the length of the park would be linked to the bridge gateway, emphasising its impressive structure and the areas created in its immediate proximity.

联合大桥是进入加莱罗滨河公园的主要入口。新的路线沿公园纵向展开，并与联合大桥相连，以突显其令人印象深刻的结构和对于整个区域的枢纽作用。

阿德莱德动物园
Adelaide Zoo

LOCATION:
Australia
AREA:
2000 m²

项目地点：
澳大利亚
面积：
2000 平方米

The Adelaide Zoo Entrance Precinct breathes new life into a once-neglected part of the City of Adelaide in Australia. Dispensing with the traditional boundary between the Zoo and its surrounds, the new entrance invites visitors to view the sights and sounds of the Zoo from public forecourts. The Zoo boasts Australia's first purpose-designed "green roof" to support wildlife shelter and extensive "living walls" of native plants, making it a significant horticultural park and research center as well as a world class zoo.

阿德莱德市位于澳大利亚。阿德莱德动物园入口区为城市里这个一度遭人冷落的地方注入了新的生机。动物园入口区的设计，摒弃了传统的横亘在动物园和周围环境之间的界限，而是让游客从外面的公共空地就能一窥园中风景。阿德莱德公园拥有澳大利亚第一个专门设计的"绿色屋顶"，既为野生动物提供避难所，也给本土植物创造了除"栖息墙"外的更多空间。"绿色屋顶"的设计不但让阿德莱德动物园跻身于世界级公园的行列，也让它成为了一个重要的园艺公园和研究中心。

The result of an ambitious integration of physical, cultural and organisational strategies, the Adelaide Zoo Entrance Precinct was designed around the core drivers of conservation, environment, education, and research.

阿德莱德动物园入口区的设计初衷是为了满足交流、环保、教育和科研的目的。它也是物理、文化和组织策略三者融合的雄心之举。

The Entrance Precinct comprises a series of interlinked forecourts that unfold over 2,000 square metres to create a natural transition and physical connection between the roads, parklands and waterways. These new links through the forecourts provide access to cafes and exhibitions via safe, lit pathways, and remediate a once unsafe part of Botanic Park – demonstrating the transformative capacity of urban design to promote safe, healthy and liveable cities.

动物园入口区外是一块互相交织、2000 多平方米的空地。这块空地是园外道路、公用场地和水路到动物园的自然过渡。如果想从园外空地进到园中的咖啡馆和展馆，游人可以走安全、装设路灯的小路，也可以绕过修缮过的植物园。植物园先前存在很多不安全因素，现已修缮完毕，充分显示出城市设计在改善安全状况、提升健康和宜居方面的革新能力。

The precinct supports a range of cultural events within the public forecourts. The 300 square metre Santos Conservation Center includes a flexible exhibition space that is accessible from the forecourts, a 100-seat theatrette, amenities and information and orientation services. The forecourts are also designed to encourage community markets and conservation industry events.

入口区前的公共空地可以举办各类文化活动。占地 300 平方米的桑托斯保护中心里有一个灵活的展览区，与公共空地相连；中心里还有一个可容纳 100 个席位的小剧院，各类便利设施及信息和方向服务。公共空地的另一个设计意图是推动社区市场的发展，鼓励受保护的产业项目。

Landscape and built form for the project have been considered as a single interwoven environment to create a unique Australian civic space. The external colour palette and materials for the precinct reflect the Australian landscape incorporating charcoal, spotted gum timber and native plants.

在设计上，希望让景观与建筑模式独特地互相交织，营造一种独一无二的澳大利亚风格城市空间。建筑在外部建材和色彩上融合了木炭、带斑点的木材和本土植物的特点，彰显了澳大利亚独特的景观特点。

奥林匹克雕塑公园
Olympic Sculpture Park

LOCATION:
Seattle Washington USA
SIZE:
9 acres

项目地点:
美国 华盛顿州 西雅图市
面积:
9 英亩

The Gardens of the Olympic Sculpture Park are the clothes that dress a 3.6 ha brownfield site on an abandoned fuel storage facility on Seattle's waterfront. A reverse "Z" shaped central path traverses roads and railroad tracks to connect urban Seattle with the vast landscape of Puget Sound and the Olympic Mountains. The project won an ASLA honor award in 2007. The park is essentially a series gardens – a metaphor for the distinctive prototypical landscapes found in the Pacific Northwest, all distinctive landscapes of a "mountain to saltwater" narrative.

奥林匹克雕塑公园中的园林占地共 3.6 公顷,位于西雅图滨海一块被废弃的燃料存储基地,土壤受污染,为棕红色。一个倒转的"Z"字形中央道路沟通了公路与铁路,将西雅图市区和普吉湾及奥林匹克山脉的宏伟景致联系在一起。此项目荣获 2007 年度 ASLA 奖。公园的核心实际上是一系列园林,这些园林展现了西北太平洋地区独特而典型的景观,是从山麓到海洋的一个地貌过渡。

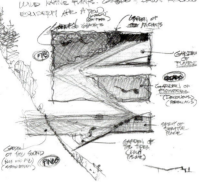

Each garden is a quintessential ecology of the Pacific Northwest which accepts succession, adaptation and evolution within an urban environment. 152, 910.7m³ of excavation material and old growth topsoil was salvaged to provide a familiar growing medium for over 85,000 transplanted native plants. A combination of soil depth, asphalt and concrete serves to cap the contaminated soil of this brownfield site.

每个园林都彰显出西北太平洋地区典型的生态类型——一种在城市环境中秉承连续、适应和改善的生态模式。为了给 85 000 多种原生植物创造一个习惯的生长媒介，我们保留了 152 910.7 立方米的挖掘材料和原有土层。我们综合利用土壤层的深度、沥青和水泥，过滤掉灰质土壤中的有害物质。

贝尼多姆市海滨公园
Paseo Maríimo de la Playa Poniente

COLLABORATORS:
Luca Cerullo—Direccidn de Obra Juan Calvo Estructura
LOCATION:
Benidorm, Spain
AREA:
40,000m²
AUTHORS:
Office of Architecture in Barcelona
DEVELOPER:
Generalitat Valenciana—Ayuntament de Benidorm
PHOTOGRAPHER:
Alejo Bague

合作团队：
Luca Cerullo — Direccidn de Obra Juan Calvo Estructura

项目位置：
西班牙 贝尼多姆市

项目面积：
40,000m²

设计单位：
巴塞罗那建筑事务所

开发商：
Generalitat Valenciana — Ayuntament de Benidorm

摄影：
Alejo Bague

Benidorm is perhaps the first Spanish town in which massive industry of leisure and tourism has turned into a paradigm; namely, that of an extremely high density concentrated in a small territory. In the long run this model has shown itself to be more efficient than others that posit the deterioration of huge tracts of land, towns empty for nine months a year, almost impossible to maintain.

贝尼多姆市土地面积狭小，建筑密集，拥有大量的休闲旅游产业，也许是最具代表性的西班牙城市之一。这种城市模式比那些建立在大片土地上、低密度、建筑物空置时间长达 9 个月的城镇模式更具效率。

In the competition for the new Promenade of the 1.5 km. along the West Beach, we propose a radical innovation in terms of what different promenades the world over have hitherto been. Not only will a borderline of protection, a hinge between town and sea, be built, but the construction will be addressed of a public place that is conducive to many different activities.

在这次 1.5km 长的西海滨长廊改造设计竞赛中，设计师提出了一个尚未运用于世界其他海滨长廊设计的开创性概念——这条长廊不仅仅是保护带，还是连接城镇与滨海地区的枢纽，它还将成为人们进行各种活动的公共场所。

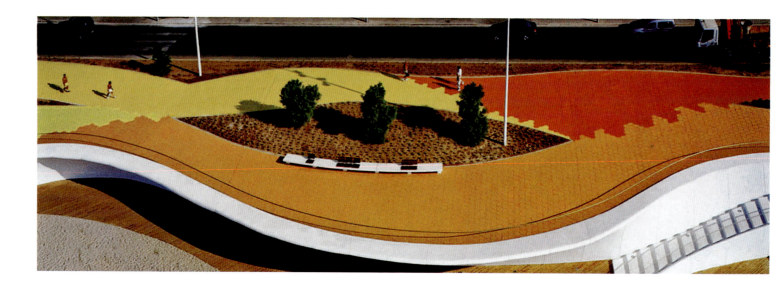

The new esplanade reduces the urbanised surface and constitutes a complex strip of transition between skyscrapers and beach.

The promenade, understood as a place with a life of its own, with organic lines, a reminder of natural wave forms, generates an ensemble of honeycombed surfaces that juggle with light and shadow, convexities and concavities that gradually construct a set of platforms and levels which will make their use feasible as areas for play, meeting, leisure or contemplation.

这里是一处富有生命力的地方，它运用有机线条，勾起人们对自然波浪的记忆，并采用蜂窝结构表面有效利用光影。凹凸相间的结构为人们提供了一系列可供娱乐、会议、休闲或冥想的平台。

查普尔特佩克公园
Chapultepec Park

LOCATION:
Mexico City

AREA:
3 480 m²

DESIGN UNITS:
Grupo de Diseño Urbano

项目地点：
墨西哥城

面积：
3 480 平方米

设计单位：
Grupo de Diseño Urbano

The major interventions in the rehabilitation of the second phase of Chapultepec Park are: Tamayo Park (12 hm²) and Gandhi Park (8.2 hm².) As outlined in the original Master Plan, the intention was to reinhabit and attract families and users to this otherwise under use area. The strategy was through specifc and discrete interventions such as: a fountain promenade (250 meters long by 20 meters wide) connecting the National Museum of Anthropology with the Tamayo Contemporary Art Museum, creating a new pedestrian axis-"paseo" through the park.

查普尔特佩克公园复原项目二期的主要工程包括：塔玛约公园（12 公顷）和甘地公园（8.2 公顷）。根据最初的总体规划，其意图是吸引居民和用户回迁到本地区，否则这里就利用不足。所采取的策略是建立一些具体、独立的设施，如：修建一个喷泉步行区（250 米长，20 米宽）将国家人类学博物馆和塔玛约当代艺术博物馆连接起来，创建一条横穿公园的新的行人散步道。

The fountain cascades and water travel through existing trees were incorporated as part of the simple design geometry. Additional comfortable seating designed by the authors was incorporated to the promenade, creating a wonderful urban oasis of rest and enjoyment.

喷泉瀑布和穿越现有树林的涉水旅行组成了场地简单几何形状的一部分。设计师额外设计的舒适座椅摆设于散步广场之中，创造了一片奇妙的城市绿洲，供人们休息和娱乐。

Additional interventions include a picnic pergola to attract families to the park, next to a new children's play area.

附加的设施包括新儿童游乐区旁边的一个野餐藤架，是一个吸引全家人集体游园的好去处。

A general simple system of paths was designed in the park, together with tree restoration, pruning and cleaning. New lighting and artistic lighting are arranged along the fountain promenade. The reception and use of the park has been dramatically increased. The Fountain Promenade has became a new identity in this otherwise neutral area of Chapultepec Park. Tamayo Museum entrance has increased substantially since the re-opening of the park.

公园内设计了一个综合的简单路径系统，和树木养护相结合，供修剪和清洁之用。喷泉散步广场周边设置了新的照明和艺术照明设施。公园的客流量和利用率急剧增加。喷泉散步广场已经变成了原本毫无色彩的查普尔特佩克公园地区的一个新标识。公园重新开放之后，塔玛约博物馆的客流量也随之大幅度上升。

Chapultepec Park is Mexico City's main public space, park and cultural complex. In 2004 a group of citizens, intellectuals and business men and women led a movement to restore and rehabilitate the famous park, together with City and Park authorities. Our office was selected to conduct a Master Plan and established 32 districts with specific plans, activities and details. They ranged from major infrastructural interventions to tree clearing, pruning and services and restoration. Two new projects were introduced by our office: a Botanical Garden and a new Fountain Promenade, connecting the famous Museum of Anthropology with the Tamayo Contemporary Art Museum.

公园是墨西哥城主要的公共场所、公园和文化综合体。2004 年，一批市民、知识分子、商业人士同城市和公园管理当局组织了一场恢复这座著名公园的运动。我公司为公园进行总体规划，并在具体计划、活动和详细计划中划分出 32 个地段，包括主要基础设施、树木采伐、修剪以及服务和修复工作。我公司引进了两个新项目：一个植物园和一个新的喷泉广场，将著名的人类学博物馆与 Tamayo 当代美术馆连接起来。

大学山镇中心和湿地公园
University Hill Town Center & Wetland Park

CLIENT:
MAB Corporation
LOCATION:
Plenty Road, Bundoora Victoria (Melway Ref 10B8)

客户：
MAB 集团

项目地点：
澳大利亚维多利亚州本多拉 丰裕路

In addition to our role as the site master planners and urban designers, HASSELL was also retained to design and document the first stage of the town center and wetland. The town center comprises the public realm treatment for the main retail street footpaths and town square. A simple and robust palette of materials including bluestone crazy paving, asphalt, concrete and granitic sand were used for ground surfaces; walls and furniture were constructed in concrete, steel and timber.

HASSELL 不仅为项目提供整体规划，做城区设计，我们也为镇中心和湿地做一期设计和出图。镇中心包括主商业街道路上的公共休闲设施和城镇广场。路面简洁，由不同的材料铺设而成，这些材料包括青石板、柏油、水泥、花岗岩以及沙子。墙壁和设施则是用水泥、钢材和木材建造的。

The use of LED lights in the seats and walls as well as the catenary lighting over the town square, provides night time interest and surprise.

座椅、墙壁和广场上空悬吊的照明设备都采用 LED 照明，让夜晚时光更曼妙有趣。

The wetland is a critical part of the stormwater strategy for University Hill. It also provides significant park amenity allowing for future connection to the Plenty River Trail and open space system. The wetland also provides significant biodiversity opportunities for University Hill (utilising an indigenous plant palette) and is a key part of the net gain strategy for the development.

湿地是大学山雨水管理系统的一个关键环节。它也是公园里的重要一景，将公园设施和裕河航道及开放空间结合在一起。湿地里生长着众多本土植物，保证了大学山地区的生物多样性，也是地区开发的净收益战略的重要一环。

HASSELL and Dryden Design collaborated on streetscape furniture design.

街景设施的设计工作由 HASSELL 和 Dryden 设计工作室共同完成。

浮动花园——永宁河公园
The Floating Gardens — Yongning River Park

CLIENT:
Zhejiang, China
TEAM:
Peking University Graduate School of Landscape Architecture and Turenscape, Beijing, China

项目地点：
中国浙江省

团队：
北京大学景观设计学研究院和土人景观，北京，中国

The park is composed of two layers: the natural matrix overlapped with the human matrix—the floating gardens. The natural matrix is composed of wetland and natural vegetation designed for the natural processes of flooding and native habitats. Above this natural matrix, float the gardens of humanity composed of a designed tree matrix, a path network, and a matrix of story boxes.

公园由两个层次组成：自然层和人造层；两个层次相互重叠，构成了漂浮花园。自然层由一个湿地和一个专为灌溉和本地生境设计的天然植被组成。人性化的花园漂浮在自然层之上，由树木、小路和故事盒组成。

合肥政务文化主题公园
Hefei Administratire Cultural Theme Park

LOCATION:
Hefei Anhui New Administrative Culture District
LANDSCAPE AREA:
50 000 m²

工程地点:
安徽省 合肥市 政务文化新区

景观面积:
50 000 平方米

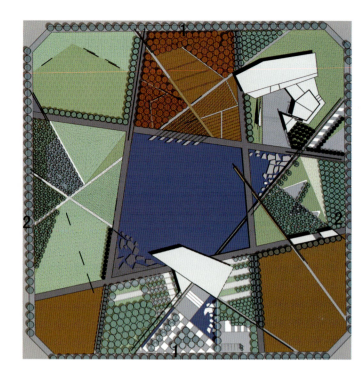

As for axis layout, we selected four main axes as park roads, then after the determination of main axes, we rotated them to an angle of 45° and forming four new ones, which overlapped with the original ones and created diverse spaces of various sizes and forms. Such design not only strengthens the connections among areas, but also made no damage to the whole axis system, which creating an affluent 3D park space effect in a simple way.

方案在轴线布局上，抽取出 4 条主轴线为园路，在 4 条主轴线确定的基础上，将其旋转 45 度角，再与原轴线叠加，形成了各种大小不一，形态各异的空间，既丰富了轴线间各区域的相互联系，又不破坏整体的轴线系统，达到了以简单的形式构成丰富的三维园林空间的效果。

In the layout of sub-areas, the main fundamental key of the whole park is green, combining with sub-area layout, and complementing with the most fundamental colors in painting, three-primary colors—red, yellow and blue, to emphasize and embody the theme—Art.

整体公园以绿色为主基调，结合分区布局，配以在绘画艺术中最基本的色彩三原色——红、黄、蓝，来突出和渲染此公园的主题——"艺术"。

As for topographical design, the design lays stress on integrity and modern sense. Large-scale and generous landscaping methods, complementing with other view elements such as large area geometric grass lopes, feature tree array and feature walls with sense of order to embody a simple, spectacular and integrated landscape result. Taking the center pond as an instance, it draw lessons from the famous method of Hongcun Village center pond, in which water pond is the center and various architectures scatter around it with inverted reflections in water.

在地形设计上，方案构成味道浓重的特征，主要强调整体性和现代感，多采用大尺度、大手笔的造景手法，如中心水池采用规则式的手法，主要借鉴了徽州著名的宏村中心水域的处理手法，水池于中，建筑物分居四周，各建筑的倒影在水中交相辉映；再配以大面积的几何形草坡、景观树阵和序列感的景墙等景观要素，均体现了一种简约、大气、浑然一体的景观效果。

Thanks to the divisions of it, the park is quite flexible in function allocation where most areas can provide corresponding capacity for different activities. We tried to create a peaceful and tranquil place and create a lightspot in Administrative Cultural New Zone where landscaping and architectures combine perfectly, all kinds of arts are home to, and itself is a piece of real landscaping arts on the earth.

由于园中主要采取分区的手法，故在功能配置上具有较大的灵活性，大多数空间都可为不同的活动形式提供相应的承载力。方案力求创造出一处平和宁静的场所，形成政务文化新区的一个亮点，在这里，景观与建筑浑然一体，是绘画艺术的家园，而公园本身亦是一幅挥洒于大地之上的景观艺术。

雷克维免下车公园
Lakeway Drive Park

LOCATION:
Narla Road Swanbourne WA
CLIENT/ DEVELOPER:
Town of Claremont

项目地点：
西澳大利亚 斯温伯恩市 娜拉路
客户/开发商：
克莱蒙镇

Lakeway is a small-scale, premium residential development of 39 lots in the leafy coastal suburb of Swanbourne, 8km west of Perth, WA. It is surrounded by Lake Claremont, a Bush Forever Site, a primary school, recreational grounds and existing residences.

斯温伯恩距离西澳大利亚州著名的旅游城市佩斯西侧 **8000m**，雷克维住宅区就坐落于这座小城绿树成荫的沿海市郊。雷克维是一个小规模的高端住宅区，有 **39** 栋单元楼。住宅区周围是克莱尔蒙湖、灌木林、一所小学校、运动场和现有的居民区。

HASSELL were appointed to provide Landscape Architectural and Town Planning services to sensitively blend the desires of the community with the ecological needs of the site. The proposal included setting aside a third of the site for bush regeneration and developing sensitive parkland interfaces to the adjoining school, residential area and a Bush Forever reserve. The subdivision layout was planned around the existing Rottnest Tea Trees and Eucalypts to retain the treed character of the surrounding neighbourhood. The site retains many existing trees and the original drive-in screen footings, which have been kept to recall the site's heritage.

HASSELL 受聘提供景观建筑和城镇规划服务，巧妙地将社区的旨趣与该地的生态需求相结合。我们的计划包括留出三分之一的空地用作灌木林重建区，设计一个完整的公园界面，体现出附近的学校、居民区和灌木丛。细部布局定在现有的 Rottnest 茶树和桉树附近，以保留周边环境的树木特征。为了回溯该址的历史承袭，该址保留了众多原有林木和先前的免下车位置。

These "old" features are complemented by the new-custom detailed seating, lookout, footbridge, boardwalk and balustrade elements, a "heritage screen" artwork, feature lot walls with integrated stone and steel detailing, special road treatments and intimate bushland tracks through the bush regeneration area that are all cohesively brought together to create a well resolved and interesting place.

在旧有景观的基础上增加了新的景致——定制的座椅、瞭望亭、人行桥、木板路、栏杆设计、展现"历史沿袭"的艺术品、石块和钢结构的景观墙、特殊的道路景观以及再生灌木林里新开辟的小路，这些因素集合在一起，给原址带来了大变化，让它更生动有趣。

霍奇米尔科生态公园
Xochimilco Ecological Park

LOCATION:
Xochimilco MexicoCity
CLIENT:
Federal District Department,Xochimilco Delegation
SURFACE:
667.17 acres

项目地点:
霍奇米尔科 墨西哥城
客户:
Federal District Department，Xochimilco Delegation
面积:
667.17 英亩

1. CULTURAL AND RECREATIONAL PARK
2. PLANT AND FLOWER MARKET
3. SPORTS AND NATURAL PARK

Xochimilco, the last remnant of the ancient lacustrian life in the valley of México, was declared "world heritage site" by UNESCO in 1987.a general plan for the restoration of the District of Xochimilco was initiated in 1989, these activities encompass: ecological rescue, hydraulic and sanitary regeneration, injection of clean and treated water to the lacustrian area and technical support in agricultural production. These activities cover an area of 3,000 hectares, including 200 km of canals and lagoons and 1,100 hectares of agricultural areas that have already been restored.

霍奇米尔科区是墨西哥峡谷最后一个古代湖上居民的生活遗址，于 1987 年被联合国教科文组织评为"世界自然与文化遗产"。1989 年，霍奇米尔科区启动了生态拯救和重建项目的总规划，该规划包括：生态拯救、水力和卫生设施重建，在湖上地区引入清洁水和再生水以及提供农业生产技术支持。这些项目辐射面积 3000 公顷，其中有 200 千米的运河和泻湖，一个重建后占地 1100 公顷的农业区。

Complementary to these actions, 280 hectares have been designated for a multipurpose park, called Xochimilco Ecological Park. The general purpose of this park is to provide Mexico City with a large natural, botanical, historical, cultural and recreational park with supporting activities and a plant and flower market, all these within an ecological park with vast lacustrian spaces and green open areas.

除上述项目外，还设计了一个面积为 280 公顷的多功能公园——霍奇米尔科生态公园。兴建此公园是为了在墨西哥市里建一个占地广、拥有较大湖上面积和绿色开放区域，集自然、生态、历史、文化和娱乐休闲功能于一体的生态公园。霍奇米尔科生态公园里不但有各种各样的辅助性活动，还有一个庄稼和花卉市场。

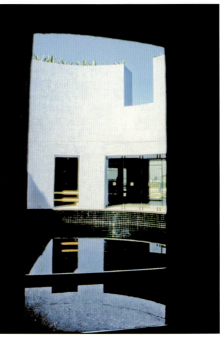

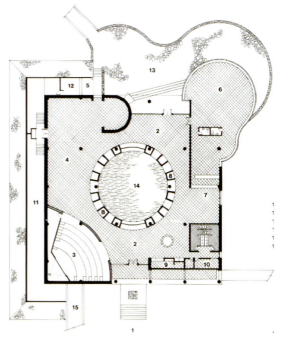

帕丁顿水库公园
Paddington Reservoir Park

LOCATION:
Paddington NSW
AREA:
4 200 m²

项目地点:
加新南威尔士 帕丁顿
占地面积:
4,200 平方米

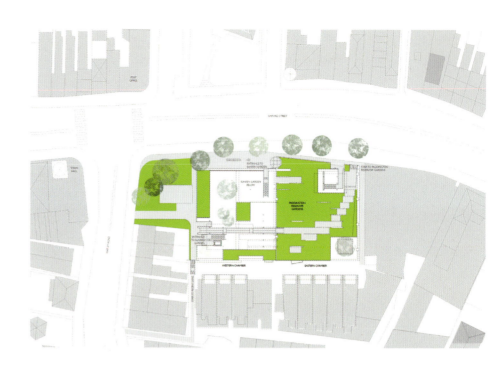

When TZG and JMD were commissioned to convert the Paddington Reservoir into an urban park, the general expectation was that the site would be capped off and a brand new arrangement built on to. However, we were captivated by the possibilities of revealing the 19th century structures as which members of the public could wander, taking in the dramatic spaces and play of light a remnants of historic walls and vaults.

当 TZG 建筑师事务所和 JMD 设计事务所接到把帕丁顿水库改造成一座城市公园的委托时，人们普遍认为水库将被掩盖起来，以便地面上会出现一个崭新的布局。然而，这座 19 世纪的构筑物最后作为一处迷人的景点展现在世人面前，人们可以周游在历史上遗留下来的墙壁和拱门之间，享受这里丰富的戏剧空间和光影变化。

Listed as a site of state heritage significance, the Paddington Reservoir was originally constructed completed in 1866 and 1878. The water chambers were built below street level with a grass opened to the public in the 1930's. The operational life of the reservoir ceased in 1899 and then as a workshop and garage until 1990 when roof collapses forced its closure.

作为国家级历史遗迹，帕丁顿水库最初于 1866 年开工，并在 1878 年竣工。水库水室位置低于街面，顶部铺设有草坪，于 20 世纪 30 年代开始对公众开放。水库在 1899 年不再提供服务，改作车间和车库，直至 1990 年，水库由于屋顶整体垮塌而完全关闭。

通州滨海公园
Tongzhou Coastal Park

LOCATION:
Tongzhou，Beijing
SITE AREA:
1,700,000m²

项目地点：
北京 通州
占地面积：
1 700 000m²

Guided by the main design concept "one river with two sides, four kinds of time-space forms, six open theme spaces, multiple scenic corridors", by using "culture line"— a ten-thousand-year old walkway, "blue line"— water source, river, "green line"— waterside scenic leisure walkway, the waterside landscape is built at multiple levels and directions with rich content.

该设计以"一河两岸、四种时空形态、六大主题开放空间、多条景观通廊"为主要思路，通过"文脉"——千年步道，"蓝脉"——水源、溪流，"绿脉"——滨水景观休闲步道，从多层次、多方位、多内容缔造完美的滨水景观。

维多利亚公园
Victoria Park

LOCATION:
Sydney Australia
SITE AREA:
24 ha

项目地点：
澳大利亚 悉尼
占地面积：
24 公顷

The landscape design concept embodies four key principles that relate to its time and place: site-wide water management strategy; interpretation of water in the new development and in the natural wetland systems; site connectivity; and community development.

景观设计彰显出四个核心原则：水域管理战略、水源在新发展阶段和自然湿地系统中的再阐释、地域的联通和社区发展。这四项原则体现了因地因时制宜。

The street layout provides simple and legible connectivity via a grid through the site with the east-west streets containing bio-retention swales in medians to collect runoff and perform initial pollution trapping. North-south streets are more traditional avenues. Utilitarian structures within the swales are combined with elegant bridge structures and a strong palette of native and endemic species to deliver a wonderful sense of enclosure and detail, greatly appreciated at the pedestrian pace.

街道的布局清晰简单，东西向街道中间设有生态洼地，收集地表径流，并率先起到治理污染的作用。南北向的街道多为传统的林荫道。实用的洼地设计与漂亮的桥梁、和周围丰富的本土物种相结合，产生了一种完美的封闭感，从步行街望去，其景致美不胜收。

The project exceeded expectations by virtue of its innovative and ecologically sustainable water management system. The bio-retention swale infiltration system regulates the quality of first flush water from the site's public roads. The filtered water is intercepted, recycled and exposed to view at the site's notable water features at Joynton Park. Plant selection and habitat creation reintroduce natural processes and promote biodiversity. Native species are predominantly used in streets and parks to re-establish endemic communities on the site and minimise maintenance requirements. That has been seamlessly integrated into a high-quality and dense urban environment provides a benchmark for water sensitive urban design in an urban context throughout Australia.

凭借创新的、可持续的水资源管理系统，此项目取得的效果远远超出了预期。生态洼地过滤体系对来自地表道路的径流起到第一轮质量控制作用。过滤后的水经拦截、回收，最后用于 Joynton 公园中的水景，成为标志性的视觉元素。在植物的选择和植被环境上也力求提升生态多样性。街道和公园里绝大部分是土生植物，目的是重建本地被破坏的生物群，减少维护的需要。地方性生物群落与高质量、高密度的城市之间天衣无缝的结合，使其成为整个澳大利亚水敏城市环境设计行业的典范。

福尔堡 Zorgpark 水道公园
Parklaan Zorgpark Voorburg

LOCATION:
Vught Netherlands
AREA:
42 000 m²
LANDSCAPE ARCHITECTS:
Buro Lubbers
DESIGN COMPANY:
Buro Lubbers landscape architecture and urban design

项目地点：
荷兰 Vught
占地面积：
42 000 平方米
景观设计师：
Buro Lubbers
设计公司：
Buro Lubbers landscape architecture and urban design

Buro Lubbers firstly formulated a master plan for the Care Park Voorburg and gradually got more involved in the landscaping and designing of the park, such as the Park lane. The goal was to develop a spatial framework, responding to new demands and needs of mental health care as well as with respect to the historic cultural property of the estate. Flexibility and attention to different target groups were key elements.

Buro Lubbers 最初是为福尔堡疗养园做整体规划，后来逐渐参与到景观和公园道路等项目的设计中。设计宗旨乃在于创造一个宽敞开阔的空间结构，在汲取宅基地深厚的历史文化底蕴基础上，能够满足精神问题疗养的新需求。公园的核心要素是注重灵活性和关注不同目的群体的需求。

10m   50m

The scenic layout of Voorburg is carefully tailored to the existing estate with its alleys, fields, water, borders, links and park areas. Sometimes existing landscape elements, such as the old beech and oak avenues, are reinforced. Sometimes new landscape elements are added such as the pond on Park lane. All land developments are, however, based on three main axes: a facility axis, a residential axis, a connection axis. As the main carriers of the landscape framework, these axes function as stepping stones for smaller operations and plans that are later realized. The facility axis, now known as Park lane, is one of the first parts realised from the master plan.

精心的设计让福尔堡宜人的风景、阡陌街巷、山水田园、园林院落与现有的建筑相得益彰。有时原有的自然景致，比如那些古老的山毛榉和橡树大街陡然增色；有时又会有新的风景加入，如园径旁新修的池塘。无论哪种，所有的景致变化都遵照以下三条轴线：以设施为轴，以住宅为轴，以关联为轴。日后的任何小规划和小改动都需依此基准，循序而行。

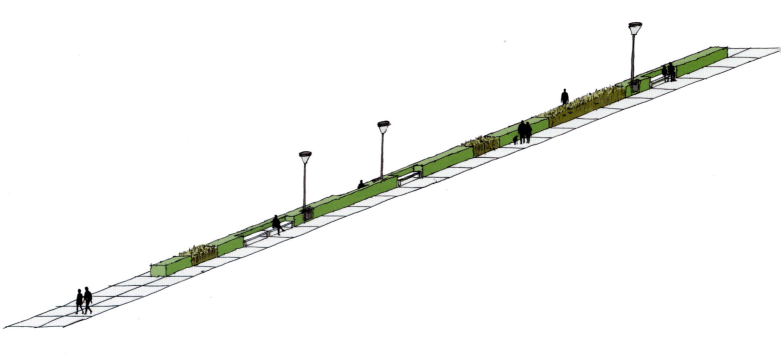

Park lane is an extension of the main entrance of the care park care. On this road facilities are concentrated, both for clients of Voorburg and residents of the town of Vught. A square with trees, a long pond, several seats on the water provide a pleasant environment for residence and meeting.

公园是疗养园主入口的延伸。这条路上的设施更为密集，既方便了远道而来的客人也为本地赫特镇的居民提供便利。临水是一个广场，遍植绿树，邻为池塘，水上几把长椅，广场既适合居民休憩也可用于集会。

核心样本花园
Core Sample Garden

LOCATION:
Québec Canada
ALL DRAWINGS:
North Design Office
IMAGES:
North Design Office, Jardins de Métis

项目地点:
加拿大 魁北克
所有图纸:
North Design Office
图片:
North Design Office, Jardins de Métis

The "Core Sample" garden was transformed for the 8th edition of the International Garden Festival, in Grand-Métis, Quebec. The garden intensifies its connection to the region through a collection of textures from the local landscape. The collection process is accompanied by a classification system to lend clarity to the new objects and their relations. Organized in a grid the core samples produce the effect of a planted field, illuminating the purpose of the core sample to determine land viability. Furthering this theme, rolling mounds flow with a variety of grains to add dynamic movement to this feature. The long grasses catch and respond to the wind, providing animation to the land forms. Suggesting multiple scales, "Core Sample" draws parallels to evoke a connection to the region and the grains that compose it.

在加拿大魁北克省 Grand-Métis 举办的第七届国际园艺节上，核心样本花园初次问世。设计师将勘探方面的系统性程序和历史引入对景观的理解上。设计师在高低起伏的地形上建造一个结构网，网里是各种核心样本，展现出公园的趋向。核心样本公园表明了公园组织和衡量内部设施的趋势，也彰显了收集、取样和发现的理念。

自由公园
The Freedom Park

LOCATION:
Salvokop South Africa
LANDSCAPE ARCHITECT:
GREENinc as part of NBGM (NLA Bagale
GREENinc MOMO)
ARCHITECT:
Office of Collaborative Architects (OCA)

项目地点：
南非 Salvokop

景观设计师：
GREENinc（NBGM 的分部）

设计师：
联合建筑师事务所（OCA）

The Freedom Park is a project mandated by President Nelson Mandela as the natural outcome of the Truth and Reconciliation Commission process that occurred after the fall of Apartheid. Its vision is structured around four key ideas: reconciliation, nation building, freedom of people and humanity. The making of the landscape seeks to recognise the spiritual origins of these ideas, and manifest them symbolically in physical form.

自由公园是在南非种族隔离暴动后，由总统纳尔逊·曼德拉授权，真相与和解委员主持的一个项目。公园的设计体现了四个核心理念：和解、国家建筑、人民自由和人道主义。公园希望能够通过景观体现出上述理念的思想源泉，在物理结构上展现其象征意义。

The Freedom Park fulfils the cultural role of a Garden of Remembrance – a natural indigenous garden telling the story of South Africa's progression to freedom. It is intended as a natural symbol for reparation, a symbol of healing, a symbol of cleansing and a place where the souls of those who lost their lives in the quest for freedom can rest. It is also a place of pilgrimage, renewal and hope for all South Africans and mankind.

文化意义上的自由公园是一座追思园，讲述着南非人民追求自由的顽强精神。追思园也象征着修复、宽谅和净化，这里安息着为追求自由而献身的先烈英灵。到追思园亦是一场朝圣之旅，这里展现了南非人民和整个人类的进步和希冀。

The Freedom Park, situated on Salvokop in Tshwane, was conceived as a narrative, a "journey to freedom" informed by traditional African culture and Indigenous Knowledge Systems (IKS) that have not been acknowledged through past knowledge or records. Five key elements: hapo, Isivivane, S'khumbuto, Moshate, and Tiva form the basis of this narrative and are linked by a wheelchair friendly pathway system that winds its way up the hill. These elements have been constructed over a number of phases and the entire project is completed by mid 2011.

自由公园位于茨瓦内的 Salvokop。建园初衷是讲述一段"向往自由之路"，讲述那些鲜为人知、少有记载的故事，由非洲传统文化和本地知识系统 (IKS) 提供文献资料。"自由之路"共分为五个部分：hapo, Isivivane, S'khumbuto, Moshate 和 Tiva。一条沿小山盘旋而上的道路将五者连接在一起，道路有方便轮椅通行的设计。上述的设计元素都已历经数个建造周期，整个项目于 2011 年中竣工。

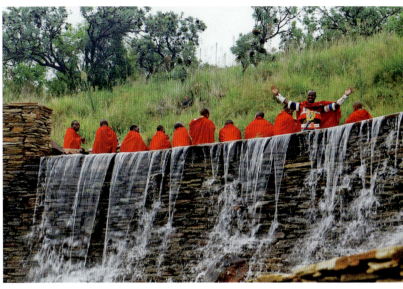

Videseter 栏杆风景公园
Videseter Railings Parkland

LOCATION:
Norway
ARCHITECTS JSA:
Jan Olav Jensen (pl), Børre Skodvin, Torunn
Golberg, AnneLise Bjerkan, Torstein Koch
AREA:
55m²
LANDSCAPE ARCHITECT:
Jensen & Skodvin

项目地点:
挪威
JSA 建筑师:
Jan Olav Jensen (pl), BØrre Skodvin, Torunn
Golberg, AnneLise Bjerkan, Torstein Koch
面积:
55 平方米
景观设计师:
Jensen & Skodvin

A scenic spot at the edge of the precipice at Videseter Falls required new railings. Annual avalanches had damaged existing railings repeatedly. The rock on which the vantage point stands had been blasted long ago to improve accessibility.

在 Videseter 瀑布的悬崖边上有一个风景秀美之地，那里需要兴建新栏杆。每年的雪崩反反复复地破坏着旧有的栏杆，那些曾经处于有利位置的岩石很久之前就已被雪崩破坏，亟须修缮。

The railings are constructed of 90 mm long steel rod, cast in holes drilled to the exact same datum in the rock. With steel plates acting as a horizontal truss, this technique resulted in a highly stable geometry. A "repairing" concrete surface has been cast on top of the rock that forms the vantage point, but the horizontal datum admits "islands" made by the ridges left by blasting. The handrails on either side of the cliff leading down to the waterfall are generated in a very different way.

在岩石的基准面上钻出精密的小洞，然后在其上浇筑上90mm长的钢柱。加钢盘做水平支撑，构成一个十分稳定的几何结构。岩石上部已经修复的水泥表面会变成有利位置，但是山脊的水平基准面仍要受雪崩的威胁。崖壁两侧通向瀑布的栏杆的修建方式截然不同。

We made principal sections of three types of railings (according to building legislation requirements) and placed the railings in plan according to the curves of the site. Then we gave some simple instructions to the welder, requiring him to build the railings on site; with the poles not closer to each other than 0.6 meters, and not further apart than 1.5 meters. They were placed on stones, the height of the vertical posts being 0.9 meters and the vertical curves of the railings forming continuous curves in conformity with the terrain directly below the railings.

根据建筑法律的规定，我们在主要部分设计了三种栏杆，而且按计划依照地点的曲线设置栏杆位置，因地制宜。然后我们给焊接工一些简单的要求，让他现场焊接栏杆，洞与洞之间最近不小于 0.6 米，最远不超过 1.5 米。在岩石上钻洞，洞的垂直高度是 0.9 米，栏杆的垂直曲线蜿蜒连绵，与地形两相辉映。

These instructions, combined with this particular process, made detailed working drawings of these quite complex three-dimensional curves redundant. The curves might actually not even be expressible when done this way, but a close likeness could have been produced had we chosen to use very advanced technology and had access to the resources needed for such a planning process. It would have required a great number of drawings, advanced programs and highly skilled architects and engineers to plan and build it. It would have been much more costly and the chances of getting things wrong during a complicated and uncertain process would have been much greater.

上述的这些步骤，加之一些特别的程序，让那些详细的施工图对于复杂的三维曲线而言很是多余。可能此种方式并不能完整地展现栏杆的曲线，但倘若我们选择高级的技术，了解计划实施的相关资源，制作一个很接近的施工图便也是可能的。我们的栏杆工程需要依靠大量的图纸、高级的处理程序和高水准的建筑师与工程师的参与来共同计划和施工。此项工程可能耗资巨大，而且在复杂、不确定的过程中出现谬误的概率也是很大的。

Gudbrandsjuvet 观景花园
Gudbrandsjuvet Viewing Park

LOCATION:
Norway
ARCHITECTS JSA:
Jan Olav Jensen (pl), Børre Skodvin, Torunn Golberg,
Torstein Koch, AnneLise Bjerkan, Sigrid Moldestad
LANDSCAPE ARCHITECT:
Jensen & Skodvin
AREA:
350m²

项目地点:
挪威
JSA 建筑师:
Jan Olav Jensen (pl)、Børre Skodvin、Torunn
Golberg、Torstein Koch、AnneLise Bjerkan、Sigrid
Moldestad
景观设计师:
Jensen & Skodvin
面积:
350 平方米

The main platform is constructed by 25mm laser cut steel sheets, cantilevered like a bridge around the cliff, hung in each end.

主观景台是用 25 毫米厚的激光切割钢板建成的。观景台悬于空中,犹如一道仪两端着地横跨悬崖的桥。

The railing has a geometry that allows it to be continuous even with very different security requirements from place to place. The large inward curve allows the tourists to securely lean out over the deadly waters. The bridges are made from different materials according to what is most appropriate at each site. The platform at the parking side is made from prefabricated elements of concrete, like a bicycle chain, an element that is connected in the corners but rotated in the angle that will fit the site.

我们采用的栏杆形式不受不同地段对安全的差异要求的影响，保证了结构的连贯性。较大的内弯线条能让游人安全地探出头俯瞰湍急的流水。根据不同地理位置的特点，观景桥的施工材料也不尽相同。停车场边的观景平台是用定制的混凝土构件建造的。构件像自行车的链条，用来连接角落，但若想在此处使用，必须旋转到合适的角度。

This was appropriate at this site because cantilevering prefabricated elements had obvious advantages economically and practically. A related geometric concept is used for the service center.

这种结构正适合用在这一地区，因为预制的悬臂构件既经济又容易操作，具有明显优势。服务中心也采用了相同的几何概念。

江苏睢宁流云水袖桥公园
Cloud and Water Bridge Park Suining Jiangsu

LOCATION:
Jiangsu China
AREA:
2 700m^2
CHIEF DESIGNER:
Yu Kongjian
DESIGN COMPANY:
Turen scape

项目地点:
中国 江苏

面积:
2700 平方米

首席设计师:
俞孔坚

设计公司:
土人景观

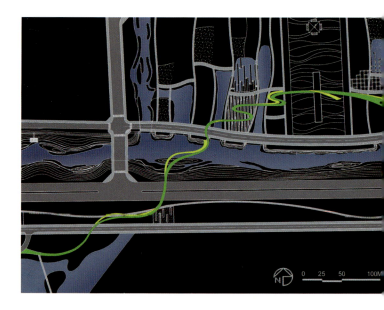

The bridge combines the complicated urban function and spatial structure together through simple and graceful municipal elements. Located in Xuning Road, Suining County, Jiangsu, the bridge passes through the express artery and many water systems, and connects the square in the core area of the county with the forest park on the opposite side of the road. The total length of the main bridge is 635m, and the total construction length is 869m, and the total area is 2,700 m^2. The total length of the four branch bridges is 242m; the width varies from 2.5m to 9m, and the gradient varies from 0.4% to 12.6%. The net height of the bridge above the road is 4.5m.

流云水袖桥通过简单而优美的市政元素，将复杂的城市功能和空间结构整合在一起。桥位于江苏睢宁县徐宁路,跨越横穿城市的快速干道和多个水系,连接县城核心区的广场和马路对面的森林公园。主桥全长 635 米,总建设长度 869 米,总面积 2700 平方米。4 道辅桥总长 242 米,桥面宽度 2.5 米至 9 米之间不等,桥面坡度 0.4% 至 12.6% 之间。流云水袖桥过路面保证净高 4.5 米。

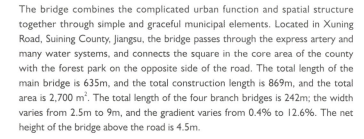

Originally the bridge is built to reinforce the connection between Hehe Square and Forest Square. It re-unifies the two open urban spaces divided by express artery -- Xuning Road and avoids the clash between cars and pedestrians when they meet on the plane surface, and guarantees the safety of the people who cross the road. Under the condition that the functional demands are satisfied, the design focuses closely on the theme of "water", which is inspired by the smooth and soft form of water sleeves. It winds and undulates in the three-dimensional space, creating the aesthetic feeling that resembles running cloud and flowing water. In addition, the elaborately designed lighting allows this bridge to dance above the urban square, water and forest more freely.

流云水袖桥最初为加强和合广场与森林广场的联系而建，使得被快速道——徐宁路划开的两大城市开放空间重新整合起来，避免了车流和人流平面相交时的冲突，保障人们的穿越安全。在满足功能需求的前提下，设计紧扣"水"这一大主题，从舞动的水袖之流畅柔美形态中获得灵感。在三维空间中婉转起伏，创造出行云流水般的美感。此外，精心的灯光设计将这条流云水袖能更自由舒畅地挥舞在城市广场、水体和林地的上空。

The bridge is the model of the perfect combination of urban landscape elements with function and form.

水袖桥是城市景观元素的功能和形式完美结合的典范。

威悉小区不来梅公园
Weser Quartier Bremen

LOCATION:
Bremen Germany
AREA:
9 800 m²
COOPERATION PARTNER:
Murphy Jahn Architects, Berlin, Chigago
DESIGN COMPANY:
RAINER SCHMIDT LANDSCAPE ARCHITECTS

项目地点：
德国 不来梅

面积：
9 800 平方米

合作伙伴：
芝加哥墨菲·扬建筑事务所柏林分所

设计公司：
赖纳·施密特景观建筑公司

The Weser Quarter is being built at the entrance to the "Überseestadt" of Bremen, directly on the Weser River promenade within walking distance of the city center. The grounds add to the coherence of the different building typologies and uses with their unified structure. On the side toward the water a terraced landscape with steps invites the visitor to take a sunbath. The terraces can be used as lounges by the neighbouring restaurants and cafés or as a grandstand for events.

威悉小区位于不来梅港口的入口处，直接建在威悉河散步区之上，步行即可到达市中心。小区协调了不同种类的建筑，采用了统一的建筑结构。在朝向港口的一边有一个台地景观，游客可以在这里享受日光浴。台地还可以用作周围餐厅和咖啡馆的休息区，或在举行大型活动时用作大看台。

社区公园景观
COMMUNITY PARK LANDSCAPE

社区公园景观
COMMUNITY PARK LANDSCAPE

圣塞巴斯蒂安 AITZ TOKI 别墅花园
Villa Aitz Toki Park in Donostia

LOCATION:
Donostia Spain
AREA:
13 500 m²
DESIGNERS:
LUR Paisajistak S.L
DESIGN COMPANY:
Paisajsitak LUR,SL

项目地点：
西班牙 圣塞巴斯蒂安
面积：
13 500 平方米
设计师：
LUR Paisajistak S.L
设计公司：
Paisajsitak LUR,SL

The villa Aitz Toki is located at the northern slope of Mount Igeldo Donostia,where you enjoy magnificent views over the Bay of Biscay. The garden is divided into two areas: the northern area overlooking the sea and the southern area.

AITZ TOKI 别墅坐落在圣塞巴斯蒂安市 IGELDO 山北面山麓，从那儿可将比斯开湾的壮丽景象尽收眼底。花园一分为二，北区可俯瞰大海，南区则是美不胜收的花园。

In the north, we just built a swimming pool with a large terrace to fully enjoy the views from this place dominating the seascape.

在北区，我们仅建造了一座游泳池，配有一个大型平台，可以饱览美不胜收的海景。

In the south, which acts as access to the house, we create a simple but powerful garden, so the garden is an attractive contribution to supplement and partly to compete with the inherent appeal of the north zone. For this purpose we choose to work with clean lines and the use of corten weathering steel as the unifying element of the whole.

南区是通往宅的路口，我们设计了一座简洁而不失美丽的花园，是另一处吸引眼球的景致，并与北区相得益彰。为了实现此设计目的，我们选择了简洁的线条，并采用了符合整个设计风格的耐候钢。

To contain the gap with the neighbouring estate, we create a retaining wall covered with corten weathering steel which spouts out a pool, also built of corten weathering steel, which empties into a linear pond built at ground level.

为了与邻近的宅邸留有距离，我们设计了挡土墙，表层为耐候钢。地面有一个线形水池，材质也是耐候钢的。

From this retaining wall emerges a planter that shelters different plantations. This set provides visual simplicity and force that contributes to complement the breathtaking views over the sea of the north zone.

从挡土墙望去，可见一个种植园，里面栽种着各种各样的植物，其景与北区令人叹为观止的海景相映成趣。

公共沙丘公园
Public Park Duinpark

LOCATION:
Velsen, The Netherlands
TOTAL PARK:
7 400 m²
DESIGNER/ARCHITECT:
Carve
TEAM:
**Elger Blitz, Mark van der Eng, Renet Korthals Altes,
Jasper van der Schaaf, Milan van der Storm**
PHOTOGRAPHER:
Carve

项目地点：
Velsen，The Netherlands

总园区：
7 400 平方米

设计师 / 建筑师：
Carve

团队：

**Elger Blitz，Mark van der Eng，Renet Korthals
Altes，Jasper van der Schaaf，Milan van der Storm**

摄影师：
Carve

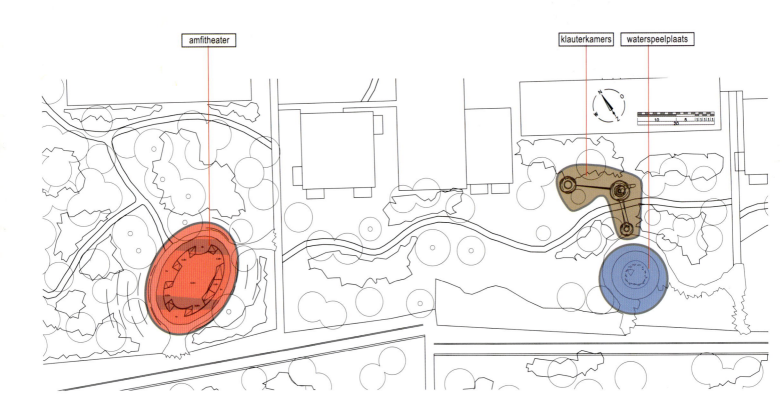

amfitheater

klauterkamers   waterspeelplaats

Our proposal consisted of three interventions. The ideas of the interventions were based on workshops done with school children living in the neighbourhood, and were sometimes additional to the already existing ideas of the landscape architect of the city of Velsen.

我们承担三个项目的设计。我们与附近学校里的小学生一起讨论，也尊重费尔森市设计师的景观设计方案。

The stage:
The stage originally planned in the preliminary plan has been turned into a multifunctional area that will appeal to youngsters all year around. By adding steel benches to an intimate area surrounded with old typical dune vegetation we introduced the skate-stage. It also gives place to open air performances as asked for by the primary schools close to the site.

高台：
高台初步定位为一个一年四季都能充分吸引青少年的多功能活动区。该区域四周是典型的沙丘植被，在旁边增设铁制长凳，将其打造成一个滑冰场。应附近小学校的要求，高台还具有举办各类户外演出的功能。

The dunetowers

Three big twisted towers, the dune-towers, are mutually connected with bridges, and cladded with diagonal beams of inland oak. As they are lowered into the surface, their immense size is easily overlooked. Once entering this secluded dune-pit they will reveal there true size. The banked sides of the pit provide an intimate place to sit and watch for children parent and family. Two towers have a double skin, one really closed made out of inland oak and one very transparent made out of welded mesh. The double skin allows for different routes through the interior of the tower, and play a game with darkness and light.

The bridges are based on the same principle one very open letting everyone experiencing the "fear" of height, and one very closed with a kaleidoscopic dazzling effect of the oak planks.

沙塔：

我们设计了三座高大、弯曲的沙塔,三者通过桥梁连通,外围是用内陆橡木做的斜纹横梁。由于沙塔部分埋在地面下,俯瞰则要比实际尺寸小得多。倘若进入沙塔之中,它们展露的形象将令人叹为观止。孩子和家长可以坐在沙塔的四周欣赏周边风景。

三塔中有两塔的外围是双层设计,里面一层是内陆橡木,外面一层则是透明的焊接网丝。双层外围设计创造了由塔内到塔外的不同线路,亮暗交替的光线变幻也非常适合嬉戏玩耍。塔间桥梁的设计基于两个因素,明显的因素是营造出"高处不胜寒"的惊吓氛围,潜藏的因素则是变化万千、让人眼花缭乱的橡木结构。

The waterplay.
The Dutch dunes near Velsen are the serve as a natural waterfilter system and are the water resource for the whole area including Amsterdam.On this place there has been a waterplay area since the sixties of the former century. We were asked to design a new waterplay area.Almost twenty aluminum masts are placed on an inclined concrete pedestal, twelve meters in diameter. Engraved with a circular equivalent of sand ripples as found on the beach and formed by the wind, the pedestals guide the water back to the recollection basin. Altogether the masts and pedestal provide for water play with surprising effects with a minimum amount of water and with a sustainable recollection of the water to prevent big quantities of suppletion water to keep the system going.

戏水
在荷兰，费尔森市附近的山丘既是天然的水过滤系统也是水源，为包括阿姆斯特丹在内的整个地区供水。20 世纪 60 年代曾在此建有一个戏水区，我们将给出一个新的戏水区设计方案。20 个直径 12 米的铝制管材铺设在一条倾斜的混凝土底座上。底座上刻有一圈圈的沙纹，像极了海风吹在沙滩上留下的层层涟漪。水流沿着底座流入蓄水盆地。在铝制管材和底座的共同作用下，最少的水能带来最大乐趣，这种可持续的用水方式也避免了浪费。

耶稣 Galíndez 公园

Jesúsz Galíndez Slope

LOCATION:
Bilbao Vizcaya (Spain)
AREA:
11 000 m²
ARCHITECTURE:
ACXT Architects
ARCHITECTS:
César Azcárate y Ana Morón
COLLABORATOR ARCHITECTS:
Xabier AparicioCarlos, Guimaraes
DESIGN COMPANY:
ACXT Architects
PHOTOGRAPHER:
Aitor Ortíz

项目地点：
西班牙 毕尔巴鄂
面积：
11 000 平方米
建筑：
ACXT 的建筑师
建筑师：
CÉSAR AZCÁRATE Y ANA MORÓN
合作建筑师：
Xabier AparicioCarlos，Guimaraes
设计公司：
ACXT Architects
摄影师：
Aitor Ortíz

To shape the embankment by using inclined planes of different materials, which reveal their strange topography to the city.

采用不同材质的倾斜面来建造护坡，以彰显城市中独特的地势结构。

The triangular planes are formed by different materials: the existing rock, vegetation of different colours, concrete in those areas which required consolidation and light, reconstructing their silhouette at night.

用各式各样的元素构筑三角形平面，包括：现有的岩石、色彩各异的植被、加固的混凝土以及在夜晚亮起构成三角形轮廓的灯管。

To create connecting elements between the top and bottom levels to lessen the impact of the embankment as a physical barrier in the city.

在护坡上部和底部之间增加过渡设计，尽量减少护坡作为城市物理屏障的影响。

Create a horizontal platform to take advantage of the height of this area of the city.

建造一个水平平台，以充分利用城市的地势高度。

All the elements in the project form part of this recreated topography: the stairways, the sitting areas, the public toilets. All these elements are included within a single fold.

此项目中的所有设计元素都得益于部分重建后的地势，因势而就：楼梯、休息区、公共厕所。任何一个独立的范围里都有这些设施。

Following the embankment project, we were commissioned to develop the Plaza de Pau Casals project and create a children's play area over a former electricity substation. All these locations are situated on steep slopes next to the embankment.

继护坡项目之后，我们受委托进行 Pau Casals 广场项目的设计，以及一个将原有的变电站改造成一个供儿童嬉戏玩耍的地方的项目。这两个项目地点都在护坡工程附近，地势较高。

半山海景公园
Mont Orchid Riverlet

LOCATION:
Velsen, The Netherlands
AREA:
7 400 m²
DESIGNER/ARCHITECT:
Carve
TEAM:
Elger Blitz, Mark van der Eng, Renet Korthals Altes,
Jasper van der Schaaf, Milan van der Storm
PHOTOGRAPHER:
Carve

项目地点：
荷兰 费尔森
AREA:
7 400 平方米
DESIGNER/ARCHITECT:
Carve
TEAM:
Elger Blitz，Mark van der Eng，Renet Korthals

Altes，Jasper van der Schaaf，Milan van der Storm
PHOTOGRAPHER:
Carve

Rolfsbukta 住宅花园
Rolfsbukta Residential Area

LOCATION:
Oslo Norway
AREA:
5 4000 m²
ARCHITECTS:
Arcasa
LANDSCAPE ARCHITECTS:
Bjørbekk & Lindheim

项目地点:
挪威 奥斯陆

面积:
5 4000 平方米

建筑设计:
阿卡斯

景观设计:
Bjørbekk & Lindheim

Rolfsbukta is a bay at the north-east end of Fornebu. It is one of the few points where residential areas on the site of the old airport will have direct contact with the sea.

ROLFSBUKTA 是 FORNEBU 东北部的一个海湾。古老海港上可以直接与大海接触的住宅区很少，而此处就是其中一个。

Very early in the project a decision was reached to reinforce the relationship of the bay with the sea and consequently a canal was extended into the bay. To bring the water as close as possible to the public we planned the inner 2/3 of the canal as a freshwater canal and the last 1/3 as a deep saltwater canal, connected by a waterfall between the two levels.

项目的早期制定了一个决策，就是加强海湾与大海之间的关系，因此修建了一条通向海湾的水渠。为了尽可能地接近水，我们计划把水渠内部的三分之二作为淡水水渠，而剩下的三分之一作为深水咸水水渠，两部分由一个瀑布连接在一起。

The residential area, Pollen, surrounding the inner part of the fresh water canal, is already completed. A large pool, a pond with stepping stones, a wooden pier and a fountain frame two sides of the Pollen buildings. The water surface is enclosed by a poured concrete ramp that slopes down on side, and on the other sides stairs lead down to a 20 centimeter deep pool. The canal is surrounded by formal beds of ornamental grasses and willow trees framed by non-corrosive steel, the same material as is used in a custom-made barbecue. There are several seats surrounding the pool made of pored concrete with wooden covering recessed into the concrete. There is also a large platform of trees planted in gravel together with long tables and benches.

围绕浅水水渠内侧而建的住宅区 Pollen 已经完成，一个巨大的水池，一个有踏脚石的池塘、一个柁墩和一个喷泉构成建筑两侧的景观。水面的一侧由一个浇筑的倾斜混凝土坡道围合，另一侧有一些台阶，通往 20 厘米深的水池。水渠周围是观赏禾草和柳树形成的整齐匀称的河床，周围采用不锈钢框架，与定做的烧烤区使用的材料相同。水池周围有一些用浇筑的混凝土建造而成的座位，木质的铺面嵌入混凝土之中。砾石中还种植了一些树木，与长桌椅一起形成一个大平台。

The seating, the "island" and the bridge are lit from below creating the impression that they are floating as the dark falls. More lighting is directed up towards the trees from the planter boxes, throughout the platforms, on the "island" and at the base of the spray nozzles of the water fountain.

座位、"小岛"和桥从下面照亮，夜幕降临时仿佛飘浮在空中一样。种植箱、平台、"小岛"和喷泉喷嘴的基座上都设置了照明，直接照向树木。

Various species of willows and cherry trees are planted close to the canal together with the ornamental grasses .

水渠附近还种植了各种柳树和樱桃树，以及一些观赏性植物。

Further out at Rolfsbukta the second part of Phase 1, Tangen and Marina, have been completed. The pier motif with promenade is an essential part of this housing complex that is made up of 6 blocks on a north-west facing slope down towards the sea. You can moor boats and walk out along the bay to the outmost tip of the bay. This sunny west-facing waterfront is designed for recreation. Poured concrete embankments provide steps and seating.

一期的第二部分，Tangen 和 Marina，位于更远处的 ROLFSBUKTA，也已完成。带有散步道的码头这一主题成为住宅综合体的基本部分，住宅综合体由六栋楼组成，位于一个朝向海面的西南向的斜坡上。你可以把船停泊在这里，沿着海湾散步，来到海湾的最远端。这个阳光充足的西向滨水区是为娱乐而设计的，浇筑的混凝土堤坝提供了台阶和座位。

丹麦王冠区花园
Danish Crown-areas Garden

LOCATION:
Aalborg Denmark
SIZE:
1 500 m²
DESIGN COMPANY:
Vibeke Rønnow Landskabsarkitekter, C. F. Møller Architects

项目地点：
丹麦 奥尔堡
面积：
1 500 平方米
设计公司：
Vibeke Rønnow Landskabsarkitekter, C. F. Møller Architects

The 12 hectar former Danish Crown abattoirs in Nr. Sundby are being transformed into a modern, mixed-use urban area, tied together by a pedestrian zone, squares and green parks. C. F. Møller Architects is responsible for the masterplan, and design guidelines, focusing on the connecting central urban "rambla" as the main axis linking the city and harbour front.

这块 12 公顷的土地曾是丹麦松比王冠区的角斗场，现在将要把它改造成一个现代的多功能城市区域，与步行街、广场和绿意盎然的公园相连。C. F. MØLLER 建筑设计事务所负责制定设计蓝图，提供设计指导。本项目旨在联通城市中央的"rambla"，使之成为沟通城市和港湾的主要轴线。

The pedestrian street is designed around a small stream, springing from the surplus of groundwater, and running through the street in an open rill to the firth. Along the way, the stream runs through mossy pools, and a larger corten steel mirror-pool. Surfaces and pavings are of honey-coloured concrete and slate, and are lit by integrated LED-lighting sources.

城中有一条小溪，成因是地下水过盛，溪水流经街面上的河道，汇入峡湾。小溪一路流过苔藓丛生的水池，经过一个较大的、澄澈如镜的耐候钢水池。步行街沿溪而建。街面和铺路建材是蜂蜜色的水泥和钢板，街道上有综合性 LED 照明设备。

毕夏普建筑区花园
De Bisschoppen

LOCATION:
Utrecht Netherlands
AREA:
500 m²
COMPLETED DATE:
2004
COOPERATION:
Köther Salman Architecten
DESIGN COMPANY:
Buro Lubbers landscape architecture and urban design

项目地点:
荷兰 乌得勒支

面积:
500 平方米

完成时间:
2004 年

合作:
Köther Salman Architecten

设计公司:
Buro Lubbers landscape architecture and urban design

De Bisschoppen (The Bishops) is a new block of buildings at the campus of Utrecht University. The architecture resulted in an inner court, two squares and several roof gardens. Buro Lubbers designed these public spaces in a user-friendly and inviting manner paying special attention to the aim of the site: a multifunctional programme of studying, working and living in a dynamic environment. The outer spaces are related by striking wooden objects that function as benches, containers for trees, cycle shed and theatre stage.

毕夏普建筑区是乌特勒支大学内一个新兴建筑区,其内包括一个球场、两个广场及几座屋顶花园。**Buro Lubbers** 是毕夏普建筑区的设计师,他将人性化和吸引力的理念融于建筑,十分重视建筑的应用目的——即在灵动的环境中营造一个集学习、工作和居住于一体的多功能建筑项目。木质长椅、树木周身的防护、木棚和演出舞台等木质产品将外部空间与建筑的内部结构完美结合,浑然天成。

0 5 10   50 meter

aanzicht plein vanaf de
Bisschopssteeg

BURO LUBBERS
landschapsarchitectuur & stedelijk ontwerp

All these objects are made from the same sort of wood and have the same finish, namely rhythmic boards of different widths. The diversified vegetation, the specific relation with the surrounding architecture and the renewed destination of relicts from the former situation, have created pleasant, social meeting places, places that students, teachers and neighbours can inhabit for several activities, from individual conversations to public manifestations.

所有木质产品均来自同一种木材，具有相同的表面特征——不同直径的树木，其纹理宽度亦不同。丰富多样的地表植被，新建筑与周围建筑的微妙关联，加之原址上旧貌换新颜的建筑共同烘托出愉悦惬意的社交活动环境，在这里学生、教师和往来行人可以随心所欲尽享乐趣，无论是私人谈话还是公众表演，毕夏普建筑区都将是不二之选。

法勒池住宅花园
Farrar Pond Residenceoppen

LOCATION:
Lincoln Massachusetts
PHOTOGRAPHER:
Charles Mayer Photography
DIMENSIONS:
805 m²

项目地点:
林肯 马萨诸塞州

摄影师:
Charles Mayer Photography

尺寸:
805 平方米

This project is situated within 12140.5m² native hardwood forest overlooking Farrar Pond, one of the many ponds linked to Walden Pond, Thoreau's historic home in Lincoln, Massachusetts. The design intent was to harmonize contemporary materials and design elements with this native plant palate and natural kettle and kame geology. A rich tapestry of native plants that transform over the seasons weaves seamlessly with the existing forest. The sculptural fence flows through openings in the forest and over various ground plane materials, both defining and blurring boundaries. The landscape design reflects the client's deep respect for land stewardship while asserting a contemporary design language that reflects the client's interest in art and sculpture.

本项目位于一片面积为 12 140.5 平方米大的天然阔叶林中，俯瞰法勒池，法拉池连通瓦尔登湖。而在马萨诸塞州林肯市瓦尔登湖曾是历史上著名的梭罗之家。设计旨在使当代材料和设计元素与乡土植物和天然凹地及沙丘地形相协调。乡土植物随季节变换呈现出多姿多彩的景象，与现有的森林相映成趣。雕刻般的栅栏穿过森林的空地来标明领地范围，又越过不同景致的地面去淡化与自然的界限。本景观设计呈现出当代的设计语言，反映了业主对艺术和雕刻的兴趣，同时体现了业主对土地管理者的深深敬意。

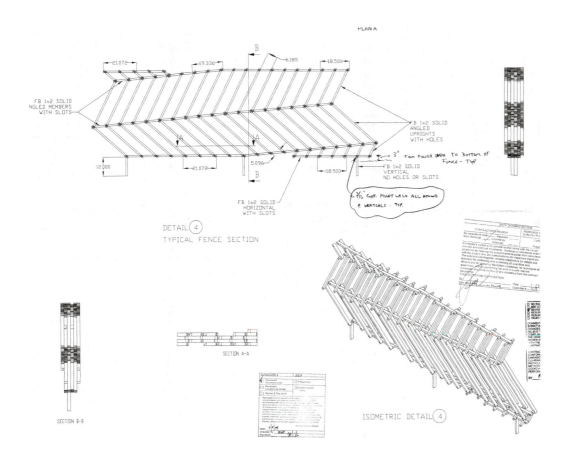

PLAN A

FB 1x2 SOLID
NGLED MEMBERS
WITH SLOTS

FB 1x2 SOLID
ANGLED
UPRIGHTS
WITH HOLES

3" FROM FINISH GRADE TO BOTTOM OF
FENCE - TYP

FB 1x2 SOLID
VERTICAL
NO HOLES OR SLOTS

FB 1x2 SOLID
HORIZONTAL
WITH SLOTS

1/16 CONT. FILLET WELD ALL AROUND
@ VERTICALS - TYP.

DETAIL 4
TYPICAL FENCE SECTION

SECTION A-A

SECTION B-B

ISOMETRIC DETAIL 4

泪珠公园
Teardrop Park

LOCATION:
New York
AREA:
0.7 hm²
DESIGN UNITS :
Michael Van Valkenburgh Associates, Inc., Landscape Architects

项目地点:
纽约

面积:
0.7 公顷

设计单位:
Michael Van Valkenburgh Associates, Inc.,

Landscape Architects

Teardrop Park transcends its small size, shady environment, and mid-block urban location through a meandering design formed with complex irregular spaces, robust plantings, strong materials and bold topography that creates places for prospect and refuge. Designed primarily as a landscape for children, the park's spatial structure and reinterpretation of natural form make a place for exploration and movement.

泪珠公园有很多不利的因素: 局促的空间、阴暗的环境以及处在道口的位置。设计师采用不规则的综合空间、充满活力的绿化、坚固的材料和大胆的地形处理打造出一个蜿蜒的步道系统,创造出景色迷人和可供栖息之所。青少年是公园的主要目标群体,因此,设计师对公园的空间结构进行了精心的设计,同时对基地原貌进行了重新诠释,形成了一个吸引人们去探索、去体验的空间。

Experiencing natural environments is widely recognized as an important part of early childhood development, and yet most urban playgrounds have banished plants in favor of equipment. Teardrop Park is designed to address this gap, offering adventure and sanctuary to urban children while engaging their minds and bodies. Site topography, interactive water fountains, natural stone, and intimately-scaled plantings contribute to an exciting inner world of intricate textures, intense scale differences, and precisely choreographed views.

体验自然环境是儿童早期开发中重要的一部分,这是大家普遍认同的一种观点,然而现在大多数城市运动场都没有采用大面积的植物,取而代之的是一些设备。泪珠公园的设计填补了这一空白,为城市中的儿童提供了探险、捉迷藏的体验,使他们的身体和智力都得到了锻炼。基地的地形、互动的喷泉、天然石以及宜人的绿化构成了一个令人兴奋的精神世界:这个世界结构复杂、富于变化,里面还有精心设计的景观。

Specific features, like the Ice-Water Wall, the Marsh with its access path scaled to children, the steeply sloped planted areas, groves of trees, and the Water Play Rocks, as well as Reading Circle where there is also an outward view to the Hudson River, celebrate the expressive potential of the natural materials of landscape construction while reinventing the idea of nature play in the city.

一些特色元素，如"冰与水"景墙、沼泽地（有一条为孩子们量身打造的小路可以通向沼泽地）、陡峭的绿化区、小树林、戏水石以及读书角（从这里可以欣赏到休斯敦河流的美丽景观），展现了天然材料在景观塑造方面的表现潜力，同时重新引进了大自然在城市中发挥着重要作用这一思想。

上海苏州河沿线住宅区公园
Shanghai Suzhou River Landscape

LOCATION:
Changning road Suzhou creek Shanghai
CLIENT:
Changning District Government.
PROGRAM:
Promenade along the Suzhou river creek, part of framework and various detailed designs for the re-development of 4.5km of Suzhou Creek edge including 5 parks.
ARCHITECTS:
Huang Fang，Li Shuyun

项目地点：
上海 长宁路（娄山关路—哈密路）
业主：
长宁区建交委
任务：
景观设计，5个公园．
设计团队：
黄芳，李淑云

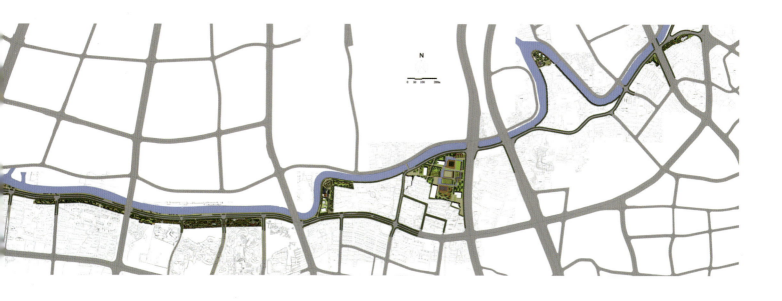

In Changning district the water edge is frequently squeezed by roads to a width of less than 5m. How can such a narrow space be construed as a leisure destination? By creating an elevated promenade, the usable space can be doubled. This sky promenade can become an attractive place for walking and playing with excellent views of the creek. It can also provide all-weather shelter to the street promenade below.

位于长宁区的河岸多数地段被车行道挤压到宽度不足 5 米。如何将这样狭长的河道建设成为一条能够吸引市民的沿河步行带是本方案首要考虑的问题。设计策略是在狭长河岸上方建设一条空中步行廊道，不仅增加了使用空间，走廊二层空间更为市民提供了观赏河岸美景的极佳视野。与此同时，一层空间则成为为游客遮风避雨之处。

The project includes the re-development of green-land between the adjacent high density housing development and the riverside road. Rather than remaining a passive buffer, can this green become an active linear park which encourages residents to get out of their enclaves and partake in a new wider community? If programs of sports, leisure and passive green are incorporated, then the park can appeal to a broad range of users. By connecting these three programs along highly legible paths, the ten 200m long parks can be understood as one 2km park. This single park can provide 2km of rubber jogging track, with periodic exercise circuit apparatus, basketball courts, badminton and other sports plazas. The park can provide 2km of socially active paths connecting tea houses, pavilions, and small plazas and gathering areas. Another 2km of quiet green experience can also be established. Where these three programmatic threads cross, or intertwine, areas of greater complexity and density emerge. To provide a place of more extreme escape, an elevated walkway is also provided. It passes the canopy of existing trees and provides a variety of views of Suzhou creek.

项目包括介于高密度住宅与河岸道路间的绿化带重新开发。除了保留其缓冲的功能，这片绿化带是否可以变成一种积极的线性公园以鼓励居民从他们的住房中出来参与到一个新的更广泛的社区中是我们想要实现的目标。如果运动、休闲和缓冲绿化带可以彼此融合，这个公园就可以吸引到大量的使用者。通过将这三个主题区域沿着道路相连接，这十条长为200米的公园就可以被理解为一条长为2000米的公园。其中一条可以提供2000米的塑胶跑道带，连接了循环使用的运动器械、篮球场、羽毛球场和其他运动广场等。另一条提供2000米长的社交主题道路，连接茶馆、临时展出馆、小广场和聚集区。第三条2000米的休闲绿色体验带也会同时建成。在这三条主题带相交或重合的地方，高密度和复杂性的区域将出现。而设计的空中步行廊道将提供一个非常自由开放的区域。它将成为经过并连接着原有树冠的天篷，为市民提供各种观赏苏州河景观的视野。

悉尼海洋生物站公园
Marine Biological Station Park Watsons Bay Sydney

LOCATION:
Camp Cove Beach, Watsons Bay Sydney Australia
LANDSCAPE ARCHITECT:
ASPECT Studios
TYPE:
Park
SCOPE:
Concept design, design development, documentation
TEAM:
ASPECT Studios Pty Ltd, Connell Wagner
CLIENT:
Sydney Harbour Federation Trust
PHOTOGRAPHY:
Simon Wood

项目地点:
悉尼 Camp Cove Beach Watsons Bay

景观设计顾问:
澳派（澳大利亚）景观规划设计工作室

设计类型:
公园设计

设计范围:
概念设计，深化设计，施工图设计

设计团队:
澳派景观设计公司，Connell Wagner 公司

甲方:
悉尼港口联邦信托基金

摄影师:
Simon Wood

The Marine Biological Station Park was formally opened on the 15th of November 2005 by Malcolm Turnbull Federal Member for Wentworth and is the recipient of a commendation award in design from the NSW AILA.

由澳派（澳大利亚）景观设计公司设计的海洋生物站公园于 2005 年 11 月 15 日正式对外开放，由 WENTWORTH 的 MALCOLM TURNBULL 联邦议员主持开幕。该设计也获得了澳大利亚新南威尔士州景观设计协会推荐奖。

The concrete slabs form an informal path to the water's edge. This beach front interface has been designed with a generous, oversized seating and stair access to Camp Cove beach.

混凝土石板形成一条自然的休闲小道通向海边。海滩设有一片宽阔、超大型的座椅和台阶，通向 **Camp Cove** 海滩。

The aesthetics of the park relies on a combination of the stunning natural surroundings and the clear simple inventions of the new park. The off-form and precast concrete interventions have both robustness and a fineness. Concrete slabs, engraved with words referencing the evolution of the site form an informal path to the water's edge. Several initiatives ensure that this is a sustainable park. The park uses native and endemic plants and is not irrigated. The design of all elements are robust and do not require ongoing regular maintenance. Care was taken to salvage and reuse as much sandstone for the walls as possible.

周边自然环境与简洁的公园设计巧妙结合，形成了公园的美学特色。现浇预制混凝土的处理方法多样，包括自然粗糙的一面，与海滩前方乡土风格保持一致；也有精美简约的一面。混凝土石板休闲小道，一直通向海边，混凝土上刻着字，展现项目现场的历史。在设计中采用多项生态措施，确保公园环境的可持续发展。例如，公园种植本土植物，不需要灌溉；设计的所有要素都是耐用、新颖的，不需要进行定期维修；维护保护区，尽可能使用砂岩来建筑墙体。

施瓦宾花园城市
Schwabing Garden City

LOCATION:
Munich Germany
AREA:
42,000 m²

项目地点:
德国 慕尼黑
面积:
42 000 平方米

The park is around 700 meters long and 70 meters wide. On the eastern side, a boulevard will invite people to walk around the way they do in a pedestrian mall. The park itself is divided into different zones, with groups of different trees and cubic pavilions; these are large pergolas with ten meters on a side. The heart of the area lies in the theme gardens, areas of relaxation for the employees lunch break. The gardens create an abstract persiflage of different landscapes that lie between Munich and the Alps. There is a rock garden, a boulder garden and a mountain lake symbolized by prism shaped structures. There is also a hill landscape marked by spherical sections, a forest garden in the form of stone pears. A meadow garden has large areas of plants. A playground invites children and grown-ups to play. Grassy areas between the gardens are available for sports and games.

公园约 700 米长, 70 米宽。公园的东侧是一条林荫大道, 人们可以在上面散步, 就像是在步行街上散步一样。公园本身分为不同的区域, 不同区域的树木和凉亭各不相同; 公园的一侧有一个十米的巨大藤架。不同主题的花园是公园的核心, 员工可以在午休的时间到这里放松一下心情。主题花园以一种有趣的方式反映了慕尼黑和阿尔卑斯山之间的不同景观, 包括: 一个岩石园、一个巨石园和一个山地湖泊, 用棱柱形的构造加以表现; 用球形元素表现的丘陵景观; 一个石头梨树构成的森林花园; 一个种植了大面积植物的草坪花园和一个供儿童玩耍、供成年人运动的运动场。花园之间的绿地可以用于运动或比赛。

Baan Sansuk 民居花园
Baan Sansuk

LOCATION:
Thailand
AREA:
11,613 m²
DESIGNER/DESIGN COMPANY :
TROP
PROJECT DIRECTOR :
Pok Kobkongsanti
PROJECT DESIGNER :
Pattarapol Jormkhanngen
PROJECT TEAM :
Wasin Muneepeerakul
Pakawat Varaphakdi
Kampon Prakobsajakul
PHOTOGRAPHER:
Pattarapol Jormkhanngen
Pok Kobkongsanti

项目地点:
泰国
面积:
11 613 平方米
设计师/设计公司:
TROP
项目总监:
Pok Kobkongsanti
项目设计:
Pattarapol Jormkhanngen
项目团队:
Wasin Muneepeerakul
Pakawat Varaphakdi
Kampon Prakobsajakul
摄影师:
Pattarapol Jormkhanngen
Pok Kobkongsanti

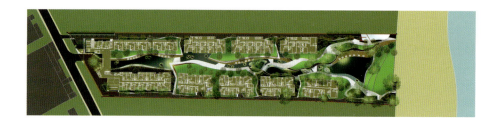

Baan Sansuk is an exclusive Residential Project, located at Hua Hin, Thailand's favorite Beach. The site is a long, noodle-like with a small narrow side connected to the beach. There are 2 rows of Buildings on both sides, leaving a long space in the middle of the site. Basically, most of the units, except the Beach-front ones, do not have any ocean View. Instead they are facing the opposite units.

Baan Sansuk 是一个独一无二的民居项目,坐落于泰国的特别受欢迎的海滩——华欣。场地是长条形的,就像面条一样,小而狭窄的一边连接着海滩。在场所的两边分别排列着两排建筑,而场所的中间则留有长条形的空地。基本上,大多数的单元,除了靠近海滩的那些之外,并不能看到海洋风光。相反,它们面向的是对面的单元房间。

Our first move is to bring "the view" into the property instead. Our Inspiration of "the View" comes from the location of the project. Hua Hin, in Thai, means Stone Head. The name comes from the natural stone boulders in its Beach area. So we proposed a series of Swimming Pools from the Lobby to the beach area, a total of 230m long.

我们行动的第一步是将"风景"带入房产。我们关于"风景"设计的灵感来源于这个项目的所在地。华欣,在泰语里意为石头。这个名字来源于坐落在华欣海滩区域的自然巨石群。所以,我们提议建设一系列的游泳池,从场所的大厅一直延伸到海滩区域,总共 230 米长。

The Pools are divided into several pools, with different functions like Reflecting Pool, Kids Pool, Transitional Pool, Jacuzzi Pool and Main Pool. At some certain area, we strategically place Natural Stone Boulders to mimic the Famed Local Beach. The result is a breath-taking Water Landscape, with different Water Characters from one end of the site to the other. These pools are not just for eye-pleasure only, but they also serve as the pools for everyone in the family.

游泳池按其不同的功能被分为几大部分,如倒映神殿于水中的石砌水池、儿童游泳池、过渡游泳池、配有极可意水流按摩浴缸的游泳池和主游泳池。在某些特定的区域,我们策略性地放置了自然石柱群去模仿当地著名的海滩景观。我们的提议带来了令人惊叹的水体景观,不同的水体特征从场所的一端变化到另一端。设置这些游泳池的目的不仅是为了带来视觉上的愉悦,同时,它们还可服务于家庭中的每位成员。

辉煌盛景公园
Parkview Eclat landscape

LANDSCAPE:
ONG&ONG Pte Ltd.
DESIGN TEAM:
Peter Frank Bridgewater
LOCATION:
Singapore
SITE AREA:
0.35 ha

景观公司：
ONG&ONG Pte Ltd.
设计团队：
Peter Frank Bridgewater
项目地点：
新加坡
项目面积：
0.35 公顷

Parkview Eclat is a high-end residential development with a water-themed landscape to complement the Art-deco style of its architectural style.

辉煌盛景（Parkview Eclat）是一个高端住宅项目，利用以水为主题的景观来实现建筑的浓郁的艺术风格。

The outdoor spaces are organised along two major axes. A main axis runs from the main building to the landscape deck, punctuated by water fountains and exquisite sculptures. This continues across the horizontal plane and even vertically up the building's façde, lending coherence to the overall Art-deco look.

住宅区的外部空间沿两条主轴线展开布局。其中一条主轴线从主楼穿过，抵达景观平台，中间有喷泉和精致的雕塑。主轴线一直贯穿整个地平面，甚至向上延伸到建筑立面，提升了与整个艺术性氛围的连贯性。

A second axis categorises various water-related activities,distinguishing facilities like the water play fountain, children pool, lap pool and Jacuzzi. The linear swimming pool acts as visual anchor along this axis, which ends dramatically in the raised Jacuzzi pavilion.

第二条主轴线包括各种类型的水上设施，如喷泉、儿童游泳池、小型健身游泳池以及极可意水流按摩浴缸。线形的游泳池与主轴线共同构成视觉景观；地势上升，主轴线最后到达地势较高的极可意水流按摩亭。

图书在版编目(CIP)数据

风景园林.1：英汉对照／郑波主编；吉典文化，千朋万友编；顾玉梅译. —大连：大连理工大学出版社，2014.8

ISBN 978-7-5611-9334-1

Ⅰ.①风… Ⅱ.①郑… ②吉… ③千… ④顾… Ⅲ.①园林设计－世界－图集 Ⅳ.①TU986.2

中国版本图书馆CIP数据核字（2014）第162210号

出版发行：大连理工大学出版社
　　　　（地址：大连市软件园路 80 号　　邮编：116023）
印　　　刷：深圳市新视线印务有限公司
幅面尺寸：220mm×300mm
印　　张：19
出版时间：2014 年 8 月第 1 版
印刷时间：2014 年 8 月第 1 次印刷
总　策　划：周群卫
责任编辑：张昕焱
封面设计：周艳丽　王志峰
责任校对：杨宇芳

书　　号：ISBN 978-7-5611-9334-1
定　　价：328.00 元

发　行：0411-84708842
传　真：0411-84701466
E-mail：12282980@qq.com
URL：http://www.dutp.cn